T0200723

Poaching and Militancy

The Asian elephant is an endangered species due to its relentless poaching mainly for ivory. However, unlike the African elephant whose both males and females are tusk bearers, in Asian elephants only males bear tusks. This has resulted in their selective killing and has not only led to an alarming fall in their number but also impacted the sex ratio.

Poaching and Militancy critically examines this problem and addresses the issue of human–elephant conflict. It studies the four elephant zones of the country with specific focus on Odisha, which is home to a large population of elephants. It also ponders on the possibility of the existence of a well-developed network supporting organised poaching and armed militancy, which applies to the central African countries as well.

Binoy Kumar Behera retired as the Director General of Police, Fire Services, Home Guards and Civil Defense, Odisha, in 2018. He received his PhD in Sociology from the Utkal University, Bhubaneswar. He was the first Commissioner of Police of Bhubaneswar–Cuttack responsible for setting up the commissionerate system of policing in the state. He was also the recipient of the Police Medal for Meritorious Service and the President's Medal for Distinguished Service.

Poaching and Militancy
The Asian Elephant under Siege

Binoy Kumar Behera

CAMBRIDGE
UNIVERSITY PRESS

CAMBRIDGE
UNIVERSITY PRESS

University Printing House, Cambridge CB2 8BS, United Kingdom

One Liberty Plaza, 20th Floor, New York, NY 10006, USA

477 Williamstown Road, Port Melbourne, vic 3207, Australia

314 to 321, 3rd Floor, Plot No.3, Splendor Forum, Jasola District Centre, New Delhi 110025, India

79 Anson Road, #06–04/06, Singapore 079906

Cambridge University Press is part of the University of Cambridge.

It furthers the University's mission by disseminating knowledge in the pursuit of education, learning and research at the highest international levels of excellence.

www.cambridge.org
Information on this title: www.cambridge.org/9781108473651

© Binoy Kumar Behera 2018

First published 2018

Printed in India by Thomson Press India Ltd.

A catalogue record for this publication is available from the British Library

ISBN 978-1-108-47365-1 Hardback

Contents

Photographic Plates

Figures

Preface

There is an African legend that elephants know exactly when they are going to die. When the time comes, they go to a place called the Elephant Graveyard where they quietly wait for death. Though no such elephant graveyard has ever been found, the emotional manner in which elephants fondle bones and particularly skulls of dead elephants they come across may have given birth to the myth that they have communion with the dead. In India too, there is a mythical aura surrounding the elephant in a tradition where the God of Wisdom is endowed with an elephant's head and auspicious ceremonies begin with an invocation to Him. However, if elephants actually had the kind of prescience they are said to have, the existential predicament now confronting them could perhaps have been less real.

An extract from an award-winning writer's autobiographical work has been included in the introductory chapter of this book where the capture of wild elephants in the jungles of Odisha's princely state of Kalahandi in the 1920s is recounted. Interestingly, when a particular herd was taken out of the forest, other elephants left. How these other herds knew of what had happened without coming in contact with those who were captured can only be surmised. Elephants are believed to communicate through low frequency sounds inaudible to human ears which, unlike their high pitch trumpeting that we hear, carry long distances. Apparently, they communicate through ground vibrations, as the soles of their feet pick up signals from the stomping of elephants elsewhere, conveying the state of agitation or otherwise they are in. However, unlike earlier times when herds could back off to the safety of contiguous forests even far away, there are not many forests left these days for them to move away to. Despite their naturally endowed ability and instinct for survival, elephants today are dying in large numbers because of their frailty against the devious ingenuity of humans and the incapability of other humans to do enough to protect them.

The study makes an effort to examine the causes of elephant death and the ramifications of deliberate killings, preventable deaths having become an increasing cause for concern. While poaching of tusker elephants for ivory has been the major area of focus, a phenomenon more recently observed has been referred to as collateral killing for want of a more appropriate expression. Collateral killing entails situations where elephants may not be the intended targets, but become victims when arrangements made to kill other animals cause elephants to die instead. For instance, live electric wire or poisoned eatables laid out for sambar and wild boar may cause elephants to die if they access the same. Again, while tuskers may be the targets, females or juvenile elephants sometimes become unintended prey when they get there first.

Even though it is not within the scope of this book, there has been a focus on tigers in India too, because the loss of different forms of wildlife is interlinked and cannot be viewed in isolation. The population of tigers falling to an all-time low is one of the possible reasons why this happened, and whether corrective measures are adequate despite the recent claims of resurgence in their numbers have been examined at some length. There has been reference to the other heritage animal of the country as well—the rhino. The study has also drawn considerably on the recent spate of elephant poaching in Africa to assess whether any lessons can be drawn from the situation, particularly in the light of reports of links between poaching and terrorism there.

The study neither claims to be a comprehensive research on the subject nor offers foolproof solutions. While examining the causes and facets of elephant killings, as indeed of tigers and rhinos, it looks at some salient points that perhaps need attention but have not been identified or adequately focussed on as yet. It tries to raise some questions on what may be going wrong in the fight against organised poaching and whether any collateral damage thereby caused could be doing more harm than good in the larger perspective. An important aspect of this has been to examine whether circumstances might have been inadvertently created to allow poachers and militants to get together to the detriment of both wildlife and security, as has happened in countries of central Africa. Even while trying to address the issues at hand, the study may have left some unanswered or inadequately answered questions for further research.

In the course of this work, the author has sometimes referred to ideas and opinions articulated by knowledgeable authorities to substantiate certain points he has tried to put across. Such references have been duly acknowledged in the book. If anything has been unintentionally overlooked, the author offers his apologies for the same. The author has also been guided by numerous years of watching wildlife documentaries on television channels like *Animal Planet*, *National Geographic*, and *Wild Discovery*. However, his viewpoints, based on three decades of close observation of wildlife around the country and its management in Odisha, that has been home to both elephants and tigers, are largely his own. This study has been done in the course of some 35 years as a senior police and government officer and in this tenure, he gained some understanding of the limitations of the institutions in the country.

The author wishes to thank his family members for their support in this endeavour. Several photographs used in the work including the one on the cover page have been taken by his younger son, Arnav. Elder son, Bibhav, a finalist of BBC's wildlife photographer of the year contest some years back, helped select and serialise them. Jayashree, his wife, was a source of strength and support, and it was her doctorate degree that inspired the author to pursue his own in a field close to his heart.

This study would not have been possible but for the infinite blessings and grace of hugging saint Mata Amritanandamayi, and it is to Amma that the author humbly dedicates this work.

Binoy Kumar Behera breathed his last in November 2018 when this book was in the final stages of publication. His family would like to thank the team at Cambridge University Press and Debabrata Swain for their help in ensuring the publication of this book.

1

The Asian Elephant

The earliest forms of simple cellular life, prokaryotes, began on earth about 3.6 billion years ago, the earth itself being 4.6 billion years old. Some 1.6 billion years later, complex cellular life, eukaryotes, came into existence and multi-cellular organisms a billion years thence. Simple animals came around 600 million years back, fish and protoamphibians around 500 million years ago, amphibians around 360 million years ago, and reptiles 300 million years back.

Mammals, the category to which both man and elephant belong, came into existence some 200 million years back. Of the ancestry of these two species, the mammoth—from which the elephant has directly descended—appeared 4.8 million years back, while humans—as they anatomically and physically appear—came only 2,00,000 years ago.[1]

In this constant process of evolution, different species have evolved over time through more specialised and efficient bodily attributes to adapt better to their living environment. In the struggle for survival, those species that could not adapt or were slow in adapting to their changing environment were marked out by nature for extinction, sometimes sooner and at times later.

The early forms of life began in water and in course of time developed features that made them more adaptable to their environment. When some moved over to land in due course, the gills or equivalent organs that they used for drawing dissolved oxygen from water gradually transformed into lungs to enable them to breathe directly from the atmosphere. Likewise, when living on land became difficult for some species, sometimes on account of excessive body weight that was difficult to carry around, they moved into water, where its buoyancy helped their mobility. Not surprisingly, the blue whale, the biggest mammal on earth that needs to breathe directly from the atmosphere, is an aquatic animal.

As the need for limbs, like hands and feet or forelegs and hind legs, was not felt in water as it was on land, such limbs gradually disappeared or were modified to form flippers and fins to help in the new living environment. While the internal system of some animals like sharks and crocodiles was already so efficient that there was little need for further evolution, others needed time—thousands of years—to evolve in order to survive in and adapt to the slowly changing environment.

With the rapacious intervention of humans in nature and its course, the process of environmental change, which has since the beginning of time been a relatively slow phenomenon, has been hastened alarmingly in recent times. This speeded-up change is not allowing many species the time to cope up with the need to evolve and adapt to the requirements of this new and fast changing environment.

The dodo, a species that has come to be regarded as the symbol of extinction, provides an illustrative example. This bird thrived in Mauritius, but things changed dramatically after humans landed in the sixteenth century and started colonising the island. This flightless bird had no known enemies until then. Continuing to live and move around as uninhibitedly as it had always done and unable to fly away to safety, it was wantonly killed for its meat. While it was not allowed the time to even begin considering humans as an enemy species until too late, rats and other vermin that came in the ships and spread around the island targeted the bird's eggs for food. From a stage when it had no enemies to one where there were so many in such a short span of time, the bird that laid just one egg a year and which did not have the camouflage to protect it in the open did not have a chance to evolve its defence mechanism and, as a result, it disappeared from the earth.

Elephas maximus

Humans and elephants have coexisted together since the age of their ancestors—the cave man and the mammoth. Stone Age humans did hunt the mammoth for its meat and woolly coat to cover themselves in the cold. However, these could hardly have caused the mammoth's extinction even though theories suggest that hunting might have been an added factor along with climate change that led to its disappearance.

In the time of the modern man, the conflict, or more appropriately, the interaction between humans and elephants, has largely been centred on man taming the elephant for his military and domestic requirements. Since the Asian elephant was more easily domesticated than the African, the two species having branched away from a common ancestor, such needs were met primarily by the former. However, the allure of ivory that has stoked the desire of humans over the centuries, and more specifically, their increasing greed over the last century, has given the conflict a different dimension.

Earlier, the population of elephants was such that the number that was captured or killed for ivory was not so much that could make a serious dent on its demography. Besides, the extent of foliage and food available in the vast forests around the country was adequate to ensure that a direct confrontation between wild elephants and humans was infrequent. That is probably why the elephant as a species did not ever face the need for evolving or developing itself in any significant way, particularly with regard to its relatively inefficient body processes.

The elephant, the largest land mammal, is handicapped by a rather wasteful digestive system that allows its body to assimilate only around half the food it eats. Given the size and body mass of the animal, it needs to feed up to 16 hours a day—two-thirds of its life time—to be able to meet its nourishment needs. For this kind of food requirement ranging from grass, leaves, and twigs to branches and bark, it has to cover considerable distances of forest area that can provide it.

Nature may have had its own reasons for not equipping the elephant with a more efficient system, possibly to sustain the innumerable creatures and organisms that live of the half-digested remains of its dung. But the harsh fact is that the animal has not been given adequate time to evolve a better system to allow greater assimilation of what it eats, which is probably what would have happened if it had had the time. In the backdrop of food supplies diminishing in rapidly dwindling forests, the survival of the elephant itself has consequently come to be at stake. With the exponential increase in human population causing people to clear more and more of forests for their agricultural needs, availability of fodder, required to keep elephants alive and healthy, is getting correspondingly depleted. With considerable areas that had been forests but have turned into agricultural land, conflicts between humans and elephants have arisen in areas that had been elephant territory. Such conflicts inevitably lead to loss of property of humans and of lives of elephants and humans alike.

To ensure its survival, many animals, in course of evolution, have developed their procreation system so that the high number of infant casualties that often takes place is compensated by large litters allowing a greater chance for at least some to survive. Alternatively, or sometimes additionally, they have short gestation periods that allow the species to breed frequently. Either way, this allows a better chance to at least the few infants that are able to survive the initial phase of their lives.

The tiger, a seriously threatened species, has a relatively low rate of cub survival because the mother tigress is a solitary animal and, therefore, has to work hard to provide for and protect its offspring. The need to go out hunting means that her cubs are left undefended for long durations, necessitating her having to shift them from place to place and time to time, away from possible predators. This is not always successful. Nevertheless, the tigress can have a litter of ordinarily up to four, and even if none of the cubs survive, her brief gestation period of a little over three months allows her to bear them again not long thereafter.

In the case of lions, which started living in groups for survival in the African Savannah, the lioness usually has a similar litter of up to four, but the survival rate is better than that of the tiger. This is because lions being social animals live in prides, and the cubs have more protection than solitary mothers. Even in the event of cubs dying as it happens when new lions take over the pride and kill infants sired by predecessors, lionesses is in oestrus immediately, and the gestation period being just over three months as that of the tiger, the process of procreation continues without much hindrance.

The elephant, however, to compound its inefficient digestive system, has a major drawback. It has a gestation period of about 22 months and gives birth to only one calf at a time. This means that ordinarily an elephant, which already is under stress for inadequate availability of food and a digestive system that does not allow it to make the most of what it eats, the mammal can give birth to just one calf every two years.

Given the fact that every calf that survives has to be cared for, milk-fed, and nurtured for some years, a female elephant can effectively produce a calf only once every seven or eight years or more. Consequently, with every calf that dies at birth or in the first few years of its existence, there is greater stress on the longevity of the elephant as a species. The time taken for the calf to get from the stage of the embryo to its premature death may be a lot more than that of some of the faster breeding species. Despite a longer life span of up to around 60, the number of calves an elephant can give birth to in a lifetime is considerably lower than with most animal species.

The Indian *Elephas maximus indicus* constitutes the major bulk of Asian elephants, the other two sub-species being the Sri Lankan and the Sumatran. The population of Indian elephant in India is 60 percent, and the remaining is spread across other countries of the South Asian mainland. The Sri Lankan and Sumatran elephants are almost entirely confined to their respective island homes. While the Sri Lankan variety is not particularly threatened by poaching because males are mostly tuskless, there are far too few left of the Sumatran. The Indian elephant can therefore be said to predominantly represent the Asian elephant in terms of numbers and vulnerability to the illicit ivory trade.

At the beginning of the twentieth century, an estimated two hundred thousand elephants used to be in the forests of India,[2] and at the outset of the current century, less than thirty thousand.[3] In other words, in the course of a hundred years—a small time span in the existence of both man and elephant—its population in India has fallen by 85 per cent. In the corresponding period, the human population has risen from 24 million to a whopping 1.2 billion, a fivefold increase.

Unlike the African elephant where both males and females grow tusks, it is only the male in India that does so. And given the admittedly superior quality of ivory that the Asian and particularly Indian elephant is endowed with, these males have in course of the last century been singled out for their ivory and their resultant liquidation. Much has been talked and written about big game hunting, particularly in pre-Independence India. Apart from providing the colonial masters sport, as most of those then involved chose to call it, kings of princely kingdoms went about decimating big game without a thought on posterity. A particular former king of an erstwhile princely state in central India is actually mentioned in the *Guinness Book of World Records* and in *National Geographic* for having killed the highest number of tigers—over 1,200. This number is just a wee bit less than the 1,400 odd felines that were believed to have remained in the wilds of the whole of India only some years back.

Similar statistics on killers of tusker elephants are perhaps not as forthcoming because it was the trophy of the tiger that held public imagination more and ensured greater bragging rights. However, this does not in any way take away from the fact that elimination of bull elephants took place on an enormous scale as the huge stockpiles of tusks, the expanse of the ivory trade, and the number of craftsmen and artisans who made their living off it testifies to. Wanton killing in India continued until the first official conservation measures in the form of Project Tiger that started in 1973.

Both the African elephant *Loxodonta africana* and the Asian, particularly the Indian *Elephas maximus indicus,* are killed for their tusks. The impact, however, is different. While both male and female African elephants have tusks and are killed in equal measure, their killing for ivory has a direct impact on their population but not so much on the gender ratio. In respect to the Asian elephant, however, it is only the male that bears tusks and gets singled out for killing. Apart from the demographic imbalance in the animal's sex ratio caused by such killing, a major problem caused by the systematic elimination of prime tuskers is that in the course of time the number of elephants with large tusks has dwindled and there could possibly not be many of them left to pass on their genes to subsequent generations. Consequently, with the male calves fathered over a period by bulls with small or sometimes no tusks, the chances of such offspring being genetically endowed with good tusks as adults are not very high.

It is within the realm of possibility that if those bull elephants that have tusks continue to get eliminated for their ivory, a stage may come in the not too distant future when all the bull elephants would be tuskless males, *makhanas*. There would no longer be any tuskers to pass on their genes down the line, and elephants with tusks could well be lost to us. That would be a sad day for both man and nature because it has already happened in Sri Lanka where only a small fraction of the males have tusks, the prime specimens all having been eliminated over the last century and more. The third sub-species, the Sumatran elephant, is already threatened by low numerical strength that has brought it to the brink of extinction.

While there had been an increase in the human population in the pre-Independence times as well, the demographic explosion that has occurred thereafter has been cataclysmic. This population explosion has led to man seeking newer areas for his needs of livelihood, particularly for occupations like agriculture. With existing agricultural land already owned and taken, areas that had forests are increasingly denuded for the purpose of cultivation. More significantly, in the process of development and industrialisation, more and more forested areas are being excavated for mines and minerals causing disruption in the elephants' corridors that had been there before. Even as poaching has not stopped after hunting was declared illegal, bringing restricted areas under cultivation and conducting mining operations against the laws of the land have not stopped either. In this process, considerable areas that had been forests and traditional elephant territory have lost their cover and only pockets in the form of islands have remained. Consequently, elephants no longer have long forested tracts as corridors and inevitably come in conflict with the humans who now occupy the intermittent areas that had earlier been forests.

How extensive the elephant corridors had been in the early part of the last century is illustrated by an extract from an award-winning autobiographical work that portrays the capture of elephants in the princely state of Kalahandi in Odisha in the 1920s.

> This wild elephant catching—the 'kheda'—was quite fascinating. Huge logs were planted close together in the ground to form a stockade on the inner sides of which were fixed thick and pointed iron nails. There was only one gate to this barricaded enclosure. Embedded with pointed iron pegs, the door to this

gate—opening upwards from hinges at the top—was lifted and held up by a thick rope. A watchman hidden atop a tree was perched nearby, the loose end of the rope tied in a knot next to him. The moment he cut the rope, the door fell under its own weight and shut the gate. Bundles of sugarcane, ripe paddy stalks, and succulent leafy branches were laid out along the route in this elephant inhabited jungle right up to the stockade and inside. A trained cow elephant of the Raja's would be let out into the jungle to guide a herd of wild elephants towards this enclosure. Sometimes people with fire torches, various kinds of drum-beating, and firecrackers sealed off three directions around the elephants and drove them along the desired route.

The moment they saw the cane and paddy the elephants would eat their way into the enclosure when the watchman sitting on top of the tree cut the rope with a stroke of his axe. The door fell shut. And held captive inside would be a small herd of wild elephants. Realising that they had been duped the elephants would try to push down the logs and break the stockade walls but heaving themselves against the pointed iron nails only left them with sore bloodied heads. The moment it got the signal, the Raja's trained cow elephant passed a loop of thick rope around a leg of each of the wild elephants, one after the other. When these jungle elephants were led out of the enclosure and the mahouts sitting atop the pet elephants around tried to secure them to sturdy tree trunks, a wild elephant occasionally got into a fight with the domesticated ones. But such bids for freedom did not usually succeed.

When some of these wild pachyderm herds had been lured into captivity through error of judgment more than once, other herds in the wild sensed danger and started moving away from these jungles. From Rampur to the Phulbani jungles, on to the Mahanadi's Tikarpara forests, across the river to the Rairakhol jungles, Bamanda, Baneigarh—one jungle touching another all the way up to faraway Assam.[4]

Need and Objectives of the Study

On 30 April 2016, Kenyan President Uhuru Kenyatta set ablaze the world's biggest bonfire of elephant tusks in the course of which over hundred tonnes of ivory worth $100 million on the international black market were put to fire at the Nairobi National Park. Even as he called for a total ban on the ivory trade and an end to 'murderous' trafficking, he echoed a strong sentiment that 'ivory for us is worthless unless it is on our elephants' and pledged that they would not be the Africans who stood by as their elephants were massacred to extinction. Africa is a home to between 4,50,000 and 5,00,000 elephants, but around 20,000 to 30,000 are believed to be killed each year for ivory which

sells for $1,000 a kilogram in the Asian market. As Ali Bongo, the President
of Gabon, puts it, elephants are being massacred in Central Africa. Rhinos
too are being slaughtered for their horn which fetches higher prices than gold
in the international black market.[5]

Ironically, in a continent where there are still close to half a million elephants
and where mass culling was in vogue in several countries in the 1980s and
1990s to keep a check on the unmanageable increase in their population,
there is a genuine fear now that unless strong steps are taken to protect it, the
African elephant, particularly the forest elephant sub-species, might become
extinct. Even while several countries continue to allow licensed hunting as a
means to keep the animal population steady, mass poaching of elephants and
rhinos and the skyrocketing price of ivory and rhino horn are driving them
to the brink of extinction in Africa. Credible and knowledgeable authorities
believe that dangerous links are established between poaching and organised
crime and illegal international trade in endangered species has integrated with
organised crime and militant groups worldwide in the fluid dynamics of the
prevalent socio-economic situation.

The Boston-based 'International Fund for Animal Welfare' which carried
out an investigation into the matter has opined that wildlife trade is no longer
an issue of environment alone now but a matter of state security. Since the latter
part of the last decade, wildlife crime has grown into the fourth largest branch
of illegal international trade with poaching having gone from one off killings
to wholesale massacres. There is greater reason to believe that militant groups
worldwide are increasingly turning to mass poaching to supply international
organised wildlife smuggling rings in exchange for arms that they need for
their militant activities.[6]

An undercover probe done under the aegis of the Los Angeles-based
'Elephant Action League' recently claimed to have established that the
Somali group Al-Shabab, which took responsibility for the massacre of some
67 innocent civilians at the Westgate Mall in Nairobi in September 2013,
uses proceeds from the illegal ivory trade to finance its terrorist operations.
The money trail connecting elephants massacred in Kenya was reportedly
tracked through this Al-Qaida-backed terrorist group to ivory buyers in Asia.
The returns from this trade are apparently so high that Al-Shabab pays its
mercenaries more than the Somali government pays regular Army soldiers.[7]
The same terrorist group, Al-Shabab, was also responsible for the attack
in Garissa University College of Kenya that killed 148 students in April
2015.[8]

Elephant ivory, rhino horn, and tiger body parts are driving these animals towards the edge in Asia too. With Asia being a major market for African ivory as established from investigations, it would be presumptuous to believe that the demand that exists for Asian elephant ivory can be taken lightly or wished away. The *Elephas maximus indicus*, one of the three sub-species of the Asian elephant—predominantly found in India—being endowed with qualitatively superior ivory than its cousins in Asia or Africa, would naturally be more vulnerable for clandestine operators to target in this highly profitable underground trade. Although studies have been done on the subject of human–elephant conflict in India, the situation in various states and regions of the country differing more in degree than in the nature of the problem, the important issue of elephant deaths that has been gaining increasing notoriety has perhaps not been studied as much. It is, therefore, not just desirable but necessary to have a closer scrutiny on the issue of elephant deaths, particularly in the backdrop of the increasingly lucrative ivory trade and the sinister ramifications of such trade reportedly affecting national security now. This is particularly so in the background of elephants and rhinos being killed for their tusks and horn in Africa when elephants, tigers, and rhinos are disappearing in large numbers in India too.

A former minister for Forests and Environment in the Government of India is on record having confirmed that some gangs involved in the illegal wildlife trade in India have links with terrorist organisations.[9] Running into millions of dollars and conducted the same way as the illegal drugs and arms trades are, elephants in India continue to be hunted down for ivory and tigers for their skin and bones. A former Prime Minister of India is also on record having said that the greatest threat to the security of the country in recent times has been Naxalite extremism.[10] Several studies including a comprehensive one done by the acclaimed New Delhi-based Institute of Defence Studies and Analyses have confirmed that the left-wing Naxalite militants, who are spread over a third of India's landmass, have been taking shelter mostly in dense forests and wildlife sanctuaries. It is primarily from there that these armed extremists wage their avowed war against the state.[11] Even while it is reasonable to presume that poachers operate from similar forests, it has not been adequately examined if like in Africa, links have developed between poachers and militants in India too.

While elephants continue to be poached for their tusks with firearms, arrows, and poison, newer methods of keeping them off crops and home have been causing their death through what may be referred to as collateral killing.

Leaving exposed live electric wire around crop fields particularly to keep off the depredatory wild boar has often been found to cause elephant deaths when the pachyderms come in contact before the intended targets do. Again, such live wire or poisoned eatables arranged by the poachers for tusker elephants sometimes cause females or juveniles to die instead as there is no certainty of who gets there first. Increasing frequency of accidents and ineffectiveness of enforcement measures have been contributing to the rising toll such that the efforts to combat elephant deaths in general and poaching in particular have become a matter of concern. While elephant deaths in the natural process do not impact their sex ratio because both male and female animals die in similar proportion, deliberate killings do.

Targets of poaching of the Asian elephant are invariably male tuskers. Such killing tends to distort the natural balance between males and females to a ratio where males, because of their selective elimination, become fewer in proportionate numbers. As already mentioned, discriminatory killing of tusker males leaves fewer of them to breed in the wild. As is eminently possible and has already happened with the Sri Lankan sub-species of the Asian elephant, if tuskers continue to get selectively and systematically eliminated, the prospect of elephants in the wild no longer having tusks is a frightfully real future possibility. From a time when there were so many of them around, the Asian elephant is today in the 'Red list of threatened species' of the IUCN, the International Union for Conservation of Nature.[12]

While analysing the whole issue of elephant deaths and the ramifications thereof, this study proposes to examine the hitherto unexplored and serious possibility of links between organised poaching and armed militancy in India and whether there have been repercussions already that have not been identified as yet. So far, studies on conflict and elephant issues have mostly been done by wildlife researchers, non-governmental organisations (NGOs), academicians, or wildlife personnel. Academics provide excellent analyses of issues at hand, and wildlife personnel too are knowledgeable about conservation needs.

However, there is often a tendency for such analyses and suggestions to be made from a somewhat restricted perspective where wildlife is focussed on but some of the other concurrent and socially relevant issues are not taken into consideration. Even while there have been studies done on the socially pertinent impact of animal behaviour with humans, these have been confined largely to the subject of human animal conflict alone. The need is felt to look from a slightly wider angle that includes, in its scope, an important feature which does not seem to have been identified as yet, let alone deliberated upon. It is therefore necessary to examine wildlife management not as an end in itself but as part of the larger issue that encompasses overall national interests, its security included.

The author, being a police officer with three and a half decades of service experience and 30 years of study of elephants and wildlife, hopes to look at the issue from this wider perspective. Having been closely associated with wildlife preservation over the years, he, by virtue of his professional experience, can also claim a reasonable knowledge of the security situation in the country. It is vital that wildlife preservation measures and those relating to security of the state operate in a mutually complementary manner and not exclusive of each other. It is the coordination in enforcement measures by the state apparatus or lack of it and the consequent fallout, if any, that the study endeavours to address.

Area of Study and Methodology

Megasthenes—Greek historian and Seleucid ambassador to the court of the Mauryan Empire—has mentioned in his work *Indika* that Chandragupta Maurya had an army that included 9,000 war elephants.[13] Kautilya, royal advisor to Chandragupta Maurya, states in the *Arthashastra*, his treatise on statecraft, that there existed nine Gaja Vanas or elephant forests in the country, the quality of elephants deteriorating as one went westwards. The sturdiest and best-suited elephants for military use were found in central India, primarily in the Kalinga Vana which comprised the forests of what presently constitute Odisha.[14] Today as well, two-thirds of the elephants of central India are found in the forests of Odisha.

Of the estimated 27,000–28,000 elephants in the country today, close to 2,000 are in the state of Odisha. These make up between 60 and 70 per cent of the elephants of central India. With the state having traditionally had a key elephant presence both in terms of quality and numbers, a study of the state would in considerable degree be a representative study of the situation around the country. This would be particularly so because as per figures available with Project Elephant in the Ministry of Forests and Environment of the Government of India, Odisha is among the states in the country that have had a high proportion of elephant deaths in recent years. Instances of poaching too are high. Even while conflict between elephants and humans is widespread in a fast deteriorating habitat, the menace of extremist violence in the state provides the required contours to analyse whether links could possibly have come to exist between organised poaching and armed militancy, both sociological phenomena being seen in several states around the country, Odisha included.

Before focussing on the specific subject of elephant deaths, the work will first endeavour to examine the broader issue of human–elephant conflict. The conflict envisages situations where humans are subjected to difficulties caused by elephant behaviour and conversely where elephants in turn are hounded by the humans. The various aspects of the former are proposed to be examined at some length and the issue of elephant deaths, particularly at the hands of humans—which is central to this work—zoomed upon in the backdrop of armed militancy.

Even while there already is considerable literature on the subject, this work endeavours to look at the problem of human–elephant conflict from a slightly different perspective. Rather than referring to the literature that is already available on elephant behaviour, the study shall attempt to study it afresh through illustrated case studies.

The most visible manifestation of conflict, crop-raiding by elephants will be studied through a series of sequentially taken photographs of a typical raid by a large herd. An effort will be made to understand the behaviour of male tuskers, individual members, and the entire herd at each stage of the crop-raiding sequence. There will also be an effort to identify and understand some of the other aspects of conflict. The damage caused to the houses by elephants in course of raiding them for paddy and the intoxicating rice brew *handia*, their increasingly aggressive behaviour, and frequent attacks on humans will also be examined, and an effort made to try and understand them. This will be done through the illustrated case study of a specific herd that had entered villages, broken into houses, and attacked people including the author who himself was very nearly killed in the process of studying them.

On the primary subject of elephant deaths, the various ways elephants are being killed and otherwise dying will be studied. Reference will be made to elephant deaths happening in the natural course, but the main thrust will be on the analysis of preventable deaths, particularly poaching and deliberate killing. Possible patterns and trends, predominantly relating to preventable deaths will be scrutinised over specific time periods, and trends in elephant deaths focussed on in line with the objectives of MIKE, the international body that monitors the illegal killing of elephants.

The criminal issue of poaching will be focussed on with regard to the illegal but thriving ivory trade and possible preventive and investigative solutions sought. The study will also delve on the recent spate of killings for ivory in African countries and whether the sometimes contemplated temporary lifting of the ban on ivory trade to facilitate African nations offload their stocks has given, and will continue to give an unwanted impetus to more poaching for ivory. Circumstantial evidence as might be available to connect primary source information to field situations will be analysed for possible confirmation of hypotheses.

A common saying in management parlance is that the first step in problem solution is identification of the problem. In the backdrop of the twin sociological phenomena of organised poaching on the one hand and militant extremism on the other, this study will try to examine evidence of possible links between the two and if there are situations where the woods might be getting missed for the trees—where there might be reasons to believe that operations relating to national security are lost sight of in the focus on preventing poaching. An important objective of the study will be to assess whether there is the desired level of synergy in enforcement measures or the inadequacy of it has helped catalyse the adversarial offensive. The relevant legal implications of the enforcement guidelines will be examined, and evidence evaluated in the backdrop of militants being known to reside in and operate from forests where poaching exists.

Tools of Data and Research Questions

Even while statistics are not always reliable and sometimes tend to project the actual ground situation inaccurately, they still remain the basic tools by which data are collected, collated, and analysed.

Official figures on elephant deaths as are available in the records of the Chief Wildlife Warden, Government of Odisha, Project Elephant in the Ministry of Forests and Environment, Government of India, and those provided by officials in the field will, therefore, be treated as reference points for review and analysis given the fact that such figures would not indicate less deaths than actually happening in the field. Figures of various causes of elephant deaths are proposed to be categorised year-wise over a little beyond the first decade and half of the current century. The various ways and numbers in which elephants have been dying during this period will be examined and the percentage composition studied.

Data as available will not necessarily be regarded as infallible, and where there is ground to question their authenticity or correctness, analysis will be made with proper reasoning both with regard to numbers and categorisation. As there is not much literature on specific aspects of elephant deaths in the country, even less in Odisha, statistics and analyses as authentically reported in important journals and other sources will also be used as the tools of data.

Answers to the questions made on the floor of the State Legislative Assembly by appropriate authorities and figures quoted therein will also be relied upon.

For a more comprehensive understanding of the situation, particularly in the study of trends, the 15-year period starting from the beginning of the century is proposed to be broken up into three sub-periods of 5 years each. A comparative analysis of figures during the three periods will be done and examined for a comparative view of patterns and trends. Reference will also be made to incidents in the year thereafter.

For a clearer understanding of the general state of affairs, individual forms of death are proposed to be clubbed together into four broad categories. These are deliberate killings, accidental deaths, deaths in natural course, and those for unknown reasons. Deliberate killings would include poaching, poisoning, and deliberate electrocution while accidental deaths would comprise accidents, train hits, and accidental electrocution. Natural course would mean natural deaths and death by disease while death from unknown reasons would be treated as such.

Patterns, if any, seen during each period and changing trends, as noticed, will be examined with reference to the actual ground situation and possible deductions made on what might be causing the changes and how more preventable forms can be checked. In addition, poaching cases registered by the wildlife wing over the 15-year period and thereafter will be collated with the number of preventable deaths actually happening for an analysis to ascertain if preventive steps are at par with incidents in the field. Simultaneously with elephant deaths, the issue of the dwindling tiger population in the country, particularly focussing on key elephant and tiger habitat areas of the state like the Simlipal Tiger Reserve will be analysed, possible trafficking routes of the illegal wildlife trade examined, and inferences drawn.

Some of the issues that will be focussed on are:

- a pictorial understanding of the conflict between elephants and humans through case studies;
- basic cause of elephant deaths, patterns and trends, and inferences if any that can be drawn;
- whether there is adequate synergy between enforcement agencies in resolving conflict and optimising measures to prevent deaths;
- whether poaching consists of isolated incidents or is part of systematised operations involving specific trafficking routes;
- modus operandi adopted by poachers and weapons used;
- reasons for the simultaneous fall in the tiger population in the state and around the country;

- whether poaching and armed militancy have developed links in India too as in Africa; and
- whether measures adopted by different enforcement agencies may have been at cross purposes and possibly have helped poachers and militants establish links to the detriment of their own operational effectiveness.

In addressing some of the issues, the primary tool that the author will use for analysis of the situation will however be his own experience in the field as a law enforcement officer and someone who has been studying elephants and wildlife for a considerable length of time.

Notes

1. Available at https://en.wikipedia.org/wiki/Timeline_of_the_evolutionary_history_of_life.
2. D.H. Chadwick, 1991. 'Elephants—Out of Time, Out of Space', *National Geographic* 179, no. 5 (1991): 2–49.
3. Available at https://en.wikipedia.org/wiki/Project_Elephant.
4. B. Behera, Ch IV (Chapter name cannot be retrieved)', in *Call of the Village*, 115–116 (New Delhi: Anamika Publishers, 2010).
5. *The New Indian Express*, Sunday edition, Bhubaneswar, 1 May 2016.
6. Available at www.ifaw.org/united-states/resource-centre/criminal-nature-global-security-implications-illegal-wildlife-tra-0.
7. N. Kalron and A. Crosta, 'Africa's White Gold of Jihad: al-Shabaab and Conflict Ivory', *Elephant Action League*, available at www.elephantleague.org/africas-white-gold-of-jihad-al-shabaab-and-conflicy-ivory/.
8. *Wikipedia*, available at https://en.m.wikipedia.org/wiki/Garissa_University_College_attack.
9. 'Illegal Wildlife Traders have links with Terrorists', *Times of India*, Bhubaneswar,6 July 2013.
10. *PTI*, 'Naxalism biggest threat to internal security', *The Hindu*, 24 May 2010, avaialble at m.thehindu.com/news/national/naxalism-biggest-threst-to-internal-security-manmohan/article436781.ece.
11. Available at http://www.idsa.in/system/files/BG_MaoistMovement.pdf.
12. *Wikipedia*, available at https://en.m.wikipedia.org.
13. 'Maurya Empire', *New World Encyclopedia*: www.newworldencyclopedia.org./ ; Pliny the Elder, *The Natural History*, VI, 22.4.
14. T. Trautman, *Elephants & Kings: An Environmental History*, 14; Kautilya. *Arthashastra*. 2.2.13-14, 2.2.15-16, 7.12.22-24.

2

Human–Elephant Conflict

In order to understand the reason behind elephant deaths, it would be desirable to first look at the broader issue of the intensifying human–elephant conflict. Humans and elephants have been coexisting for long, and even while humans have domesticated elephants for their needs, the two have always had separate areas for habitation. For the former, it has been open land suitable for their livelihood, and for the latter, the forests. It is not that the demarcation between the two areas of living was always so rigid that one did not infringe on the other: humans being dependent on and entering forests for many of their needs and animals often straying into areas of human habitation. However, this was more the exception than the rule and not to the extent as would cause much hindrance to either's undisturbed existence in its respective habitat. It is when this balance gets disturbed to the extent of normal habitation becoming difficult that things begin to go awry and the roots of conflict originate.

Human–elephant conflict can be broadly categorised into two kinds—one, what humans do to elephants, and the other, what elephants do to humans. The former primarily consists of humans causing elephant deaths which, being the integral subject of this study, will be examined in a separate chapter. Persecution by humans of elephants foraying into crop fields and households is a reaction to elephant depredations and hence not strictly a separate feature of the conflict. The second aspect of the conflict, which is what elephants do to humans, and what has been agitating people against them, is the raiding of crops and households as also their injuring and killing of humans. As the former trait is examined separately, it is proposed to illustrate what elephants do to humans from a slightly different perspective through two self-explanatory, personally experienced case studies.

Crop Raid

Crop raiding by elephants is both a cause and symptom of conflict. For the large number of farmers whose crops and thereby their livelihood is destroyed, it is the prime cause of conflict. From the elephant's perspective, however, habitat destruction by humans that limits its forage options is actually the root cause of conflict and crop raids by elephants, only its symptom. Whichever way one looks at it, crop raiding by elephants is the most visible and widespread manifestation of human–elephant conflict. This most distinguishable trait therefore needs to be illustrated in some detail.

Case Study 1—Crop Raiding Pattern of a Large Elephant Herd

The sequence of photographs in the following pages depicting a typical crop raid is a selection from among several taken by the author. This was in the Betnoti area of Mayurbhanj some time back during the migratory visit of a mega herd of over a hundred elephants from the Dalma sanctuary area of the state of Jharkhand into Odisha's border districts of Balasore and Mayurbhanj. This migration, coinciding with the paddy ripening and harvesting season in November–December, has since become an annual feature in the human–elephant conflict calendar and the cause of considerable farmers' agitations.

Ordinarily, mega herds like this one do not exist on a regular basis. The common need for food causes smaller herds to get together in a combined migration, and as they move on, more such herds and individual tuskers join. Bull elephants are ordinarily solitary and are not herd animals. When a male calf is born into a herd, usually headed by a matriarch, it stays on until its adulthood at the age of 12–14 years after which it is driven out and has to go on its own way. Male adults become very sexually aggressive because of the excessive secretion of testosterone in their system during seasonal musth. Driving them out of the herd they were born into is nature's way of ensuring that there is no inbreeding. Solitary tusker bulls joining the herd for the common cause of feeding, as is seen in the ensuing photographic study, is a typical example of species adapting against their general nature in their need for survival.

It is not certain whether the migration of elephants coincides with the paddy harvesting season because of the inadequacy of forage in their original habitat or for the allure of the more nutritious ripe paddy. Presumably, it is the latter. However, as it would be the paddy harvesting season near their original habitat as well, the reason for the elephants to not remain there and instead undertake this migration is because local resistance to smaller individual herds would be much more difficult to counter. The gradual formation of a mega herd is practically more convenient for the protection of its members in course and at the end of a long migration. Knowing that people will make efforts to obstruct and drive them away from places where there are crops, the elephants realise that they have a better chance of standing up to human resistance if in a large group than in small individual herds.

A distinct difference in the way a large herd such as this operates and smaller ones do is that small herds and individual tuskers on their own usually take cover of night time for raiding crops and village households. Large herds take to crops even during daylight, usually in the late afternoon, and continue into the night. However, they do not often visit households in large groups because of the limited options available in them, for many of the elephants. Nevertheless, they do break up temporarily into smaller groups for the convenience of night-time foraging and depredation of village homes. Taking advantage of the dark to make most of inadequate human resistance and then regrouping during daytime for their security needs makes eminent sense.

After return from the migration roughly around the time when the paddy season is over, the constituent smaller herds and tuskers usually go on their own way in smaller groups or on their own.

Since the mega herd has been causing considerable crop damage in the two odd months that it comes visiting and foraging, steps have been taken to keep the elephants away through erection of solar fences, digging of trenches, and use of improvised physical deterrents. They had not been able to enter in such large numbers collectively during a few paddy harvesting seasons thereafter, and they broke up into smaller groups over a wider area. But nature always has its own way of getting around situations in the process of evolution and the need for survival. Staying upwind of deterrents like chilli powder, using stretches of trench that have filled in, and traversing sections of solar fencing where power has been disrupted, they were back again. There will be new methods to keep them out of farmers' paddy land and counter-manoeuvres on the part of the elephants to get around them. In this constantly evolving cycle, what further measures humans come up with and how the elephants react and where can be known only in the times to come.

Crop Raiding Pattern of a Large Elephant Herd

A bull tusker emerges from the forest in the late afternoon. During daytime, elephants ordinarily remain under cover of the forest and come out to the crops in the fringe areas between the forest and villages as evening approaches. When raiding crops, the more confident and fearless bulls take the lead as can be seen here. The rest of the herd is still inside the forest, and many of the members at the edge of it and waiting to come out when they get the signal. Unlike this large herd which comes out of the forest when it is still light, small herds and individual bulls, if on their own, usually emerge only after it is dark.

Plate 1 A tusker bull elephant emerges from the forest

Plate 2 Two more tuskers join the first

Two more bulls join the first. Adult tuskers do not ordinarily form part of the elephant herds and are generally solitary. A male calf born into a herd is milk-fed by its mother till he is old enough to feed with the herd. On attaining adulthood, he is driven out of the herd, this being nature's way of preventing inbreeding. Male lion cubs too are allowed to stay in their pride until adulthood. However, unlike expelled young adult lions that will otherwise be killed by the lion or lions that have come to head the pride thereafter, young adult male elephants are driven out by the herd matriarch, the alpha female. Again, unlike expelled lions who lead nomadic lives till they are strong enough, single or in more numbers, to take over a pride by force, male elephants stay solitary most of their lives and get back to herd life only temporarily for mating or participating in such migrations.

Plate 3 The tuskers move towards the paddy

A closer view of the tusker bull in the lead is shown in Plate 3. The two tusker males behind are smaller in size and younger in age and experience. As is noticeably apparent, the lead tusker exudes the confidence that inspires his fellow male peers to follow him. Other members of the herd, largely females and a few subadult males, have still not gathered the confidence to come out in the open. A few can be seen in the background and through the foliage. Only a female and a subadult have come out of the forest but are still not confident enough to follow the tuskers.

Plate 4 The tuskers from another angle

As the tusker approaches the ripe paddy, the two others follow. The confidence of the first has rubbed off on the two and they are visibly more self-assured. In the background, the subadult that had earlier been on the right of the photograph in Plate 3 behind the female has now got beside her and is partly hidden. She is evidently its mother. The other female that was partly visible through the foliage is no longer seen, while the rest of the herd still remains inside the forest, not confident enough to venture out yet. As may be seen, the paddy in some of the plots has already been harvested, but in some others, it is still standing. It is where the uncut paddy is that the tuskers are headed towards.

Plate 5 The lead tusker begins feeding

Poaching and Militancy

The lead tusker starts feeding. The two others behind will follow him in a matter of moments. Taken from a different angle, the photograph does not show the location of the female and the subadult who were the only other two who had come out. The two apparently continue to wait where they had been. However, if there is no human interference and the lead tuskers settle down to feed, they and the other elephants in the forest will soon gain the confidence to join them.

Plate 6 Other herd members come out into the open

More members of the herd emerge from the forest but do not feel confident enough to follow the three tuskers to the paddy yet. The female, who with her subadult calf was the first to venture out after the tuskers, can still be seen standing in an unchanged position facing away, which suggests her nervousness. The subadult calf meanwhile is typically frisky, first was behind the mother, then partly hidden by her on the far side, and now alongside her on her left, looking in the direction of the tuskers. Even while the subadult has been constantly changing its position, it has not left the security of its mother's side.

Tuskers no. 2 and 3 can be seen on the extreme right of the photograph, while no. 1 is the bigger one seen behind hidden by the two except for his back seen above them. They have settled down to a leisurely feed, and their relaxed demeanour is a signal to the rest that there is nothing to worry about. Other members of the herd that had emerged from the forest and been waiting now move towards the crop from the left of the photograph to the right. Even while there is uncut paddy spread around the fields, they go first towards the security of the tuskers' proximity. It is only after their risk perception diminishes that they will themselves spread around to wherever the paddy is.

Plate 7 The tuskers settle down and others move towards them

Plate 8 The herd settles down, but front liners are cautious

The herd has settled down and is relatively less anxious now as can be seen from their movement away from the proximity of the tuskers. However, they have not let their guard down as can be judged from the elephants in the foreground continuing to face the local people head on in case they interrupt their feeding. Even though the villagers assembled are many, the herd was large and spread over a wide area. People realised that the elephants will retaliate from a different direction if they try to drive them away from one direction. This explains the elephants' formation of large impermanent herds in the common need for food and survival. In this mind game, people too are aware that such large herds are much more difficult to drive away.

Plate 9 The herd has settled down to a relaxed feed

The herd has now completely settled down and is relaxed. Having initially shown caution by facing the assembled villagers while feeding, the elephants have now realised that the people will not disturb or try to drive them away. They are not bothered about their presence any longer, and having lowered their guards, now they turn around in course of their eating as it is convenient to them. That they no longer see any threat from the villagers and can choose to ignore them is evident from the fact that most of them now have their backs to the people unlike the previous photograph where they were facing them.

Plate 10 No paddy for the infant yet

The migratory mega herd is a loose conglomerate of smaller herds and individual bulls that have gotten together in a common cause. There are therefore elephants of all ages and sizes. Many female members of these smaller herds have given birth in recent times, and in their midst there are several infant calves that are not old enough to feed on paddy yet. But it is dinner time and as social bonding among elephants is strong, infants too partake in the mood of mass feeding by settling down to their own diet of mothers' milk.

Plate 11 Feeding over

Whatever paddy was there has mostly been eaten and there is no reason for the herd to hang around there much longer. As it is to be late evening and with visibility already low and darkness setting, the security of forest cover is no longer needed for the place for their next visit. It is time to get going.

Plate 12 Time to move on

As it is now well into dusk, the elephants' need here is fulfilled and it is time to move on. The herd is less circumspect now that it is to be dark. It will move on to other paddy fields to feed during the night and visit the local water source for a drink before getting back to the security of the forest before daylight.

House Damage and Attacks on Humans

Increasing loss of habitat and growing number of forays by elephants into crop fields and human habitations have often united local people boisterously against their depredations. People explode fireworks and high decibel crackers near the elephants, beat drums and cymbals, and use various noise irritants like vuvuzela type of sound instruments to drive them away. There have been instances not long back in mining towns like Joda in Kendujhar district—the hotbed of mining operations where loss of forest habitat has been colossal—when people have tried to deter marauding elephants by throwing improvised Molotov cocktail types of petrol bombs at it. When this particular tusker that had been repeatedly entering the township area and causing mayhem trampled a youth to death, people who had prepared themselves with fragile glass and plastic containers filled with petrol lighted the wicks stuck into them and flung them at it. On impact, the containers broke and the petrol spilled over the elephant, spreading the fire over it.

Even while such acts of violence may not kill the elephant, they can make it vicious and more confrontational. Although elephants are known for their ability to identify and remember, violent interventions by people often cause them to become aggressive: retreating during daytime and emerging after dark often to visit human habitations and people who come in the way. Constantly repulsed by retributive locals using noise, fireworks, and other forceful means, elephants have become belligerent and abrasive.

Plate 13 Tusker fury and retaliation at Joda, the heartland of mining operations in Kendujhar District[1]

In a curious reversal of habitation character, elephants have attacked villages with such frequency and brazenness as almost to have become visiting residents there. Villagers in turn, not knowing which area will be targeted next, have been leaving home and village, even taking refuge on the branches of large trees to escape night-time pachyderm rage. The picture shown is with reference to the herd whose crop raiding has been photographically depicted earlier in this chapter. Back in Balasore and Mayurbhanj districts of Odisha from Jharkhand in greater numbers a year later, they first break up into groups and spread panic in and around the local villages at night time. Having got the people terrified, their efforts to feed become much easier.

Plate 14 In a reversal of occupancy habits, wild animals seem to have taken to villages and humans therein to trees[2]

The case study narrated below, an attempt to understand and analyse the aggressiveness of elephants, was originally written as an article based on the author's personal experience and observations on elephant behaviour over a period of time. The use of the first personal singular 'I' in the original has been retained in the narrative.

Case Study 2—Anatomy of an Elephant Attack

Having been observing and studying elephant behaviour for close to 30 years since being posted in the late 1980s as Superintendent of Police of the district of Mayurbhanj where the Simlipal Tiger Reserve is located, the varied experiences one has had in their midst have always provided new insights into their ways. Herds have been seen to behave differently when they encountered members of other herds: sometimes being aloof and unreceptive and at other times going out of their way to greet them. Ordinarily genial, alpha females with infants and calves have been seen to be belligerent in inverse proportion to the size of their herds, because smaller numbers make them feel much more vulnerable.

Aggressiveness

What causes elephants to become aggressive? Matriarchs usually allow males to stay in the herd till they reach adulthood after which the latter have to leave and fend for themselves. This makes eminent sense because it prevents inbreeding and allows the herd male's genes to spread over a wider base. It is, however, during this nurturing in herd life that young males get to be groomed in the ways of intra- and inter-species societal living which is what guides their behaviour when they are on their own later. If the early fostering has not gone well, they tend to become intolerant and confrontational.

There was this unusual phenomenon in the Pilanesberg Game Reserve in South Africa where tusk gores and crushed ribs that caused the death of several rhinos were found to have been the handiwork of teenage male elephant delinquents. These youngsters had been relocated from their culled herds when young, and having had no role models to learn from, they had not learnt to share space with fellow herbivores.[3]

Even while such instances are not very common, bull elephants can become violent and sometimes rogues because of painful experiences they might have had or more commonly from the high level of testosterone in their system during the seasonal musth breeding time. In respect of female elephants, particularly matriarchs however, the tendency to be aggressive is almost always associated with threat perception to the herd, particularly calves. In recent times, frequent confrontational situations with humans have been precipitating corresponding acts of elephants' proactive belligerence.

It is a matter of conjecture and debate if the behaviour of animals can be analysed through similar parameters of psychoanalysis as we do in respect of humans. The father of psychoanalysis, Sigmund Freud, had suggested that the physical inadequacy and the mental association that comes with it often cause individuals to become more aggressive than normal.

Does that explain why tuskless *makhna* bulls tend to be more aggressive than tusker males because—despite their usually larger size—they do not possess tusks, a potent weapon and the most identifying attribute of male Asian elephants? Incidentally, the most savage killer lions in recorded wildlife history were probably the two that killed over a hundred workmen and prevented the construction of a rail bridge at Tsavo in present-day Kenya for over a year in 1898. Interestingly, both the lions were maneless. The behaviour of two other male lions at the Maasai Mara game reserve in Kenya documented by *National Geographic* more recently offered an interesting insight into animal psychology.

The lions were at a kill they had just brought down when the sight of some Maasai tribesmen walking in their direction caused them to abandon their kill and flee. There was a time in the past where a Maasai tribal youth was not considered a man until he had killed a lion and taken its tail as trophy.

Though the last lot of lions to have faced such threat was many decades ago and generations of the felines have since been protected, something apparently told the lions that tall and confident young men carrying spears and dressed in red were to be feared.

Is animal behaviour sometimes also guided by a subconscious commonality that psychoanalyst Carl Jung called the 'collective unconscious'? This could possibly be so because all life forms are products of the same process of evolution, and man and animal are perhaps linked by psychological traits that differ in degree rather than in content.

Charging Elephants

Of the numerous times that I have been chased by elephants, solitary male or herd female, and whether I was on foot or in a vehicle, mock charges have far outnumbered genuine ones. It is possible to a considerable extent to anticipate whether an elephant will make believe a charge or is likely to seriously do so. If fidgetiness, noisy rumblings, raised trunk, and fanned out flapping ears precede or accompany the charge, it can be reasonably inferred that the animal may not actually attack. Getting the source of the threat intimidated is largely its objective, and like many others in the animal kingdom, this effort to appear larger and overaggressive is actually its defence mechanism at play. A short burst chase in supposed pursuit enhances the intimidation effect, which is precisely what the elephant wants to achieve. However, if such loud symptoms are absent and the possibility of a charge exists, there might be reason to worry.

There was an occasion to witness this in the border districts of Odisha where a large herd of elephants from the Dalma jungles of neighbouring Jharkhand had been congregating during their mass winter movement. This mega herd had actually been a conglomerate of herds that had gotten together in a common migration and several adult tusker bulls also joined them in course. Even while the bigger bulls and the larger females with calves amongst them seemed indifferent to the presence of people around, a spirited, mid-sized member of the herd was threatening to break loose. One could see that it was edgy, and when it finally did charge after several noisy demos, I lowered my camera and without moving from where I stood, held my hand up in the manner of a traffic policeman showing a 'stop' signal. It abruptly stopped, tail-hoisted halfway up on one side, and with a squeal and a fart, turned around and rejoined the herd.

Wildlife documentary makers Mike Penman and Jeff Corwin have an interesting take on what happens when an elephant expects to intimidate someone and the person does not actually get intimidated or shows that he is not. Apparently, the elephant gets confused and not knowing what to do next, usually backs off. The documentary visual of 'Mad Mike' keeping a charging African jumbo at bay by holding his ground and pointing at the animal makes interesting viewing. Jeff Corwin too has a similar visual of a defiant yell at a charging elephant forcing it to turn back. But it is not very certain if these are genuine situations that depict authentic elephant behaviour or enactments for the purpose of television viewing.

Proper perception of animal behaviour helps in better understanding of conflict situations with them. With humans increasingly coming in direct confrontation with elephants, it was desirable to know if this premise on elephant behaviour held true in situations of genuine attack, then those who faced the inevitability of such attack could use it as a last option to save themselves. There was an occasion to put the scope of such surmise to practical test in rather trying circumstances some time back.

The Makhna *and the Herd*

The state of Odisha, home to over 60 per cent of the elephants of central India, has been witnessing an increasing trend in human–elephant conflict situations, thanks to increasing deforestation and diminishing habitat caused by agricultural, industrial, and mining activities. Consequently, injuries and deaths of both humans and elephants have been steadily rising over the years. Increasing instances of extreme violence to keep elephants off house and crop have in turn provoked them into greater aggressiveness and acts of confrontation. Though a spate of incidents having been reported from the Karanjia subdivision of Mayurbhanj district and a study of such conflict being a part of my research work, I thought it would be desirable to collect photographic documentation of houses and properties damaged, people mauled, and of the herds responsible.

It was the fourth of September 2012. Portions of the walls of Suresh Tiria and Rajkumar Tiria's houses in village Kasia under the Raruan Police Station (PS) limits had been brought down by the pachyderms during the night, and paddy kept by their families was eaten. The traditional intoxicating rice brew, *handia*—which tribals in these parts of the state ferment in their households, and the smell of which draws elephants from afar—was also consumed. Even while the inmates and other villagers fled to the village outskirts, one of them came in the way of the largest member of the herd and was battling for his life, now in the local subdivision hospital.

There were reportedly two herds ravaging this area for some time but had since gone their separate ways terrorising different villages on their own. The six-member herd led by a particularly large animal that had visited this village was reportedly now in the Singda jungles not far away. Going by the description the villagers gave of the herd leader and its behaviour, it was probably a tuskless *makhna* male, as its proactive aggressiveness was not quite consistent with the defensive belligerence of herd matriarchs. Apparently, the bull had temporarily joined the herd for the purpose of feeding.

It was around four in the afternoon when we drove to the alighting point and walked the half kilometre or so to the Singda jungles where the herd that had visited the Tirias' house the previous night was. The heavy rains caused by a low pressure formation had left the fields leading up to the jungle slushy. The elephants that had been out in the open had retreated into the undergrowth as the noisy congregation of local villagers was getting them flustered.

There were four of them in the bushes, not entirely visible but present as the breaking of branches and movement in the foliage indicated, sometimes reasonably close by and sometimes some distance away. The din created by the villagers was hindering their emergence into the open, and as the late afternoon hour not leaving much daylight time for photographing them, I went to the edge of the clearing and was able to get some photographs of a partly hidden mid-sized male with lean but well-formed tusks. Its restiveness, ear flapping, grunts, and shaking of the head when I showed up indicated that it was stressed but the same symptoms also suggested that there probably was no actual threat from it.

After sometime, the big bull showed up in the bushes farther away and its back visible above the foliage. Other members of the herd, hidden in the thicket, were apparently some distance farther up. Even while elephants in jungle areas can ordinarily be detected fair distances away from the sound of breaking branches and loud rustling as they move, they can move like phantoms if they want to, furtively and soundlessly. As the small herd was obviously stressed by the constant intervention of villagers preventing them from coming out into the open and feeding on their paddy, it was necessary to be watchful. Being at the forefront and realising the importance of being vigilant, I alerted the officials immediately behind to keep a sharp eye for movement in the bushes around lest we be surprised.

The big elephant seemed to have moved farther away into the jungle and was out of sight for some time when suddenly, from an entirely different direction from where it had been, it came up between us and the subadult tusker. Taking us completely by surprise, it was when he walked with brisk paces and stopped face to face some 20 feet away that I realised that he had not made a sound. The posture was clearly menacing, and the unmoving trunk, steady head, and fixed gaze were telltale signs of trouble. After several moments of eye-locked stalemate, I started to slowly retrace my steps while continuing to face him but realised that if I slipped in the soggy earth as I was likely to walking backwards, my positional vulnerability would trigger an immediate attack. Again, there was so little distance between us that if it charged and I held my ground, the slippery soil would prevent it from stopping even if it wanted to. This would leave me defenceless and at his mercy.

And then even while backing away I did something I should perhaps not have done in the circumstances, even less because my personal well-being was at stake. But my research drive and the uniqueness of the situation got the better of me. As this was the animal that had brought down the walls of the houses I had earlier visited, that had been feeding on the paddy that the inmates had stored, the animal that had been attacking people and terrorising the locality, and in effect an animal that directly represented the species in the conflict between humans and elephants, I wanted its photograph for my records. I clicked the shutter of my camera, and even as his body language expressed disapproval and he came menacingly closer, I clicked it a second time. He charged.

Close Encounters

Some years back at the Simlipal Tiger Reserve, there was an occasion when I came across an extremely reclusive mother elephant with a calf less than a month old. I had seen similar mother and infant in small groups of three quite a few times before but never just two of them by themselves in tiger territory. On another occasion many years earlier at the same Upper Barakamuda region of southern Simlipal, I was tracking a mother, small calf, and subadult in the dark with a torch light when I was surprised by a stalking tiger, flailing tail raised right up, getting between me and the elephants. So brazen and single-minded was its pursuit of the calf that the tiger did not seem to notice my presence.

If that could happen to a family of three elephants, this tiny calf alone with its mother was even more vulnerable. Hence, she was nervous, as much of possible predators as of human intruders. Unable to get any photographs of the two during the evening light as they kept retreating, I took my chance in the moonlit night at the salt lick near the Range office. Even while it was rather dark for the camera, I was thrilled to see through night-vision glasses the increasing clarity of the head-on view of the mother.

When I lowered the glasses and suddenly realised that she was coming at me, I grasped the predicament I was in, turned around, and ran. Slowing down some 60 to 70 yards on with the presumption that she would have stopped, I found her in hot pursuit. I again sprinted a further 70 to 80 yards and was convinced that she would have given up by then because her calf was all alone. I had underestimated her maternal protective instinct. I now fled for my life and, as the elephant got nearer, turned sharply to my left, and continued to run. A series of short, ill-tempered trumpet blasts told me that the chase had ended. Until then the elephant had not made a sound and moved like a ghost in the moonlight.

I learnt two things that day. A silent elephant charging is a dangerous elephant, and if you are being pursued and the elephant has gotten close, it makes sense to abruptly change direction which the elephant cannot do as quickly because of its body momentum.

Attack

That was then. Back in the present, I had no option but to do the same as the bull elephant came after me. The gathered villagers fleeing in all directions confirmed that he was in close pursuit. Adopting the same strategy that had helped years before, I ran a short distance briskly to allow the pursuing elephant to gain momentum and as I then turned a sharp angle as planned, slipped in the sludge and fell. Sitting up, I turned around and watched in dismay as the elephant, some 50 to 60 feet away, bore straight down on me.

Elephants cannot jump, and at no stage, even when they are running, do they entirely lose contact with the earth even for a moment. Unlike other quadrupeds that leap off the ground when running, the elephant's scurrying creates the impression of gliding which does not always allow an accurate estimate of its briskness. The torso and upper body appear to be static and the only moving parts are the legs. Like all quadrupeds, they have knees and ankles, but even while the knees of the front legs do not bend as much forward, the ankles bend back and forth almost 180 degrees which creates the impression of the elephant kicking with its lower legs as it runs.

Even as all this played out before me, my heart sank when I saw its trunk steady and curled inwards, the ears held back at the sides, and its head determined and low. More ominously, the fact that it had not made a sound only heightened the sense of foreboding I already had. Getting up from the slush and running were out of the question, and he would easily catch up. And knowing I was scared and showing it, he would not leave me a chance. Now desperate, I pinned my last hopes on the belief that an elephant did indeed back off when it saw someone it was after not getting intimidated.

Sitting up straight, I raised my right hand in the direction of the approaching animal and pointed my index finger at it. Looking it directly in the eye and contorting my face to the most fearsome expression I could make, I let out as loud, prolonged, and piercing a yell as I could.

I was still staring it in the face and yelling when the elephant struck. What followed was surreal. As I rolled over under the brute force of two enormously weighty kicks and then saw the elephant's right hind leg kick me again, I drifted into a daze wondering if this was a dream or I was actually in the process of being killed.

Many thoughts passed through the mind in those few moments. Mad Mike and Jeff Corwin had got it wrong and I was a goner. How mangled would this giant animal leave me? After 25 years of elephant watching, was this how it was going to end? Joy Adamson spent an entire lifetime studying lions, and till it was proved otherwise, she was believed to have been killed by one. But the similarity of my imminent death with that of a wildlife celebrity like her was little consolation. I also remembered a sadly disfigured gentleman from West Bengal I had met many years before, a tranquilising expert who could not get out of the way of the rogue tusker that he was required to immobilise and got brutally walked over. Sprawled helplessly in the mud, I curled myself into a foetal position and waited for the bull elephant to trample me.

But the elephant did nothing of the kind. It hurried on back into the jungles, and incredibly, I pulled myself up on my own. I looked around at the several pieces that my faithful Nikon D80 was now in, and still in a daze, I actually walked half the kilometre to my vehicle. However, when within a half an hour my left calf bloated like a purple balloon, my left thigh felt paralysed and turned a deep blue, and the agonizing pain in my hip suggested pelvic and lower vertebral fracture, I returned to my headquarters strapped to a stretcher in an ambulance.

Analysis

Even while doctors—who examined the split muscles and acute haematoma in my upper and lower left leg and the hairline fracture in my back—advised prolonged treatment, it was a small price to pay for my life. The bull and his newly found herd continued breaking into houses and till last count, had killed at least three people. The list of those killed by elephants in recent times in the state of Odisha should have included my name too but the answer to why things happened differently probably lay in the tracks that the elephant left behind in the mud.

As the elephant headed straight for me, I did not have a realistic chance. The show of defiant aggression on my part, however, even though an act of desperation, had got him puzzled enough to the extent that he wanted to avoid me, and the tracks in the mud suggesting that even as he came close the bull elephant had tried to move away. Having been only 40 to 50 feet away when I raised my hand and yelled, the distance was far too less for him to entirely change direction given the slushy ground condition and the momentum of his running weight. But he did try to move away even as his left foreleg kicked me flush on my lower left leg and right on my thigh. As I was tossed to one side and he moved away on the other, his right hind leg brushed an oblique blow on my lower back, which, had it been a frontal kick, would have shattered my pelvis and lower spine.

Significantly, when my raised arm was pointing towards the elephant and inches away from him, his trunk reached for the camera hanging from my neck and not my arm. As corroborated by eye witnesses, the elephant plucked the camera with its strap over my head and flung it to the ground, breaking it to pieces. Having given vent to his anger on the object that provoked him rather than the person who used it, and having simultaneously landed his kicks, he turned around in a wide circle and ran back into the bushes with the same alacrity with which he had begun his pursuit.

My fate would have been different if the elephant's kicks had impacted bone instead of flesh and muscle, even more if he had stepped on me. But I would not even remotely have had a chance had not the years of elephant watching allowed my wits to be around when I needed them the most.

Confrontations such as this should be avoided, but when in extreme situations, it often becomes a mental test between man and elephant. The experience provides ground to believe that the psychology of elephants differs from humans perhaps in degree only. The calmer they are allowed to remain, the lesser the chances of conflict. However, when constantly flustered by humans driving them away from home and crop, and the elephants themselves not finding adequate forage elsewhere, they become feisty and irritable.

Different species represent different stages of the evolutionary process which include the mental aspect of such evolution too. Just as in intra-species confrontations, the matching of mental faculties between man and animal could sometimes decide the way things develop in inter-species conflict situations as well.[4]

Anatomy of an Elephant Attack—Case Study Illustrations

Plate 15 The mid-sizer charges

While photographing crop raiding by a large elephant herd, this fidgety mid-sizer threatened to charge after repeated noisy demos. Even while other members of the herd were not too concerned about the human presence in the vicinity, the elephant charged at the author, and the photograph was taken by his son standing nearby. The fact that it had for quite some time been giving indications of charging suggested that it was nervous and wanted to intimidate rather than actually do any damage, a manifestation of its defence mechanism at work.

Plate 16 Retreating at raised hand

When the elephant charged, the author lowered his camera, and without moving from where he stood, he held up his hand to it. The elephant braked to a halt and backed away. There is reason to believe that elephants expect those they charge to get intimidated and show it. In the event of this not happening, they sometimes get confused about what to do next and back off.

Plate 17 The elephant bull's calling card

Suresh Tiria of village Kasia under Raruan Police Station of Karanjia and his family lived in this brick-walled house where they stored paddy grown for the year from the small landholding they had. The *makhna* bull ate the corn they had grown over a small patch of land just outside. Having detected the paddy in the house and the intoxicating rice brew *handia* which he would have smelled from far, the bull pushed down the wall to get to them. As the family members ran away to save themselves, the paddy they had kept for the coming months was consumed by the raiders as was the *handia*. Consumption of the intoxicating brew usually sets off elephants into greater brazenness and uninhibited behaviour.

In the same village, Rajkumar Tiria, who maintained a rather large family of seven from his small landholding and odd jobs lived in this mud-walled house. In course of the visit to the village during the night, the bull elephant pushed down a portion of the wall and consumed the paddy that Tiria had kept for his large family's needs. The family fled and watched from a distance along with other residents of the small village as the elephants took their time, and the of the residents who inadvertently got in the way was seriously injured.

Plate 18 Another house visited

Plate 19 The bull and a herd subadult tusker

The six-member herd that raided the houses of Suresh Tiria and Rajkumar Tiria in Kasia village under Raruan Police Station of Karanjia was led by the bull on the right. Increasingly violent resistance from villagers to elephants entering their paddy fields has been making elephants more irritable and belligerent. Driven by the need for food, the herd led by the bull entered villages at night time, broke into houses, and terrorised villagers. As per reports received, he had killed at least three people and injured many.

Plate 20 The bull's surprise appearance

Having photo-documented crop raiding by elephants, the author had been trying to illustrate house breakage, the other aspect of human–elephant conflict. With not much time left in the late afternoon to photograph the bull that had been leading the rampage, the author had gone up to the edge of the forest for a photo-op. From his position earlier, the elephant had moved into the forest. After long moments of inaction as the author waited, the *makhna* tuskless bull suddenly appeared from an entirely different direction closer than anticipated. The menacing look on its countenance is unmistakable.

Having suddenly emerged from the forest close by and taken the author by surprise, the elephant stood some 20 feet away staring straight at him. The unmoving body and trunk, ears held back close to the sides of the head, and the direct, unflinching gaze were telltale warning signs. Sensing it, the author started to backtrack his steps in the slushy ground. However, not wanting to let the front on photo opportunity for his study go by, he clicked his camera a second time. This enraged the bull enough to attack.

Plate 21 Last photo clicked before the attack

Plate 22 Anger vented on the author's camera saved his life

With no way of getting up and outrunning the elephant after the author slipped and fell, he put on an act of counter-aggressiveness. Even as the puzzled animal tried to move away, the distance was far too less and he repeatedly kicked and very nearly trampled the author. In an interesting revelation of elephant behaviour on diverting the focus of its aggressiveness, the elephant did not hold the author by his outstretched hand that was inches away. Instead, he plucked the camera from around his neck and flung it to the ground. Having given vent to his anger, he ran back into the forest leaving the author badly injured but alive.

Conflict Management

With the exponential increase in human population, growing dependence on agriculture and increasing unavailability of land, forests have gradually been denuded for the purpose of cultivation, beginning with the fringe areas. Such conversion penetrates deeper into the forests, depriving elephants of the habitat they once had and leaving far too little for their sustenance. The effort to restore elephant habitat in forest land that has been lost because of cultivation is perhaps the single biggest requirement in mitigating conflict.

While the Mayurbhanj Elephant Reserve in Odisha comprising Simlipal in Mayurbhanj, Hadgarh in Kendujhar, and Kuldiha in Balasore has been in existence for quite some time and continues as such, it was proposed to expand the Sambalpur and Mahanadi Elephant Reserves to allow more habitat area to the pachyderms. There has in recent times also been the proposal to create two new reserves, namely the Baitarani Elephant Reserve and the South Odisha Elephant Reserve to allow more expanse of contiguous forests to elephants. However, the balancing line where the economic concerns of a developing state and the need to keep nature undisturbed are to be drawn has not exactly been identified. Large tracts of forested areas are mineral rich and exploitation of such resources matter in the efforts to industrialise a developing society. On the other hand, economic exploitation of these areas would further deplete the natural forage available to elephants and push them further into conflict with humans dependent on agriculture.

The existing elephant reserves in Odisha and those that are proposed to be further included could be depicted as in the map below.

1. Mayurbhanj Elephant Reserve—existing
2. Baitarani Elephant Reserve—proposed
3. Sambalpur Elephant Reserve—expanded
4. Mahanadi Elephant Reserve—expanded
5. South Odisha Elephant Reserve—proposed[5]

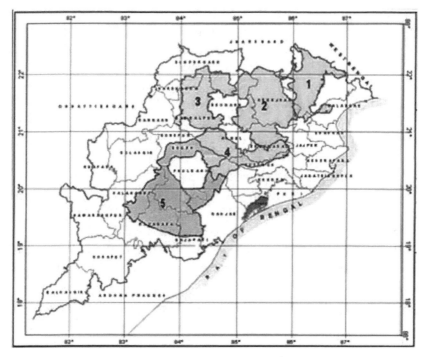

Figure 1 Existing and proposed elephant reserves in Odisha

Even while the proposal for Baitarani and South Odisha Elephant Reserves has been shelved because of the need to tap these mineral-rich areas commercially, the expansion of the Sambalpur and Mahanadi Elephant Reserves is pending consideration with the state government. Meanwhile, the proposal for establishment of 14 separate corridors for movement of elephants: three of them connecting the districts of Mayurbhanj and Kendujhar in Odisha with the state of Jharkhand, and one connecting Mayurbhanj with the state of West Bengal are proposed. The remaining 10 elephant corridors are proposed within the state covering the districts of Mayurbhanj, Balasore, Kendujhar, Dhenkanal, Angul, Nayagarh, Sambalpur, Sonepur, Boudh, Kandhamal, Rayagada, and Kalahandi.[6] Nine more corridors in the districts of Dhenkanal and Cuttack are contemplated.[7]

The need for creation of corridors has been recognised as imperative for all elephant habited states of the country. Recently in 2017, the Supreme Court of India directed the central Government to take measures for making 27 elephant corridors over nine states in the country inviolate. These projects, three of which are in the state of Odisha, had been spelled out in the report

Gajah in 2013 and accorded high priority by Project Elephant of the Ministry of Environment and Forests in the Government of India.[8]

However, in the absence of forests over a considerable stretch of the proposed corridors and presence of human settlements already there, only time will tell how far these corridors can be effectively converted for undisturbed elephant movement. Since elephants which move long distances for food cannot be forced into moving along specifically created corridors unless there is edible allure therein, it is essential that plantation of nutritious forage material like thorny bamboo and other wholesome plants preferred by elephants is adequately done along such corridors. Alternative plant feed can also be contemplated. Creation of water bodies intermittently for the drinking needs of elephants, each of whom consumes 150–200 litres a day, solar fencing along some stretches, and excavation of trenches to keep them off human habitations are also proposed. However, there continues to be debate on the efficacy and cost effectiveness particularly of trenches that require considerable investment and costly restoration maintenance after rains.

As per figures with Project Elephant of the Government of India, Odisha figures are among the highest in the country in respect of both elephant deaths and humans killed by elephants. In the first decade and half of the current century, 919 humans were killed in elephant attacks in the state at an average of 61 each year. Of these deaths, 371 have been in the last half decade at an annual average of 74, and the 99 reported in 2014–15 being the highest for a year. Two hundred and sixty humans have been injured during the period and the average each year being 17 which has risen to 28 over the last 5 years. This would mean not only that the frequency of elephant attacks on humans has been increasing in recent times but, significantly, also that only 22 per cent of the people attacked survive.

The 12,206 houses and 1,24,406 acres of crops damaged in elephant depredations during the corresponding period mean that on an average they have broken into 813 houses and ravaged 8,294 acres of crop each year. The average when seen over the last 5 years rises to 877 houses and 13,297 acres of crop which is consistent with the increasing trend in attacks on humans.[9] In other words, both the frequency and intensity of conflict are progressively increasing.

As crop raid, house damage, and attacks on humans were generally being the main areas where people are at the receiving end in the conflict with elephants, it was desirable to know from stakeholders what they thought were the areas of inconvenience and comfort. Separate questionnaires were sent to a cross section of the affected public, wildlife officials, and police officials having jurisdiction over the areas where the people often disturbed public order as a result. The responses were illuminating.

Even while the number of deaths caused by elephants has been steadily rising, it has primarily been crop damage that has got people agitated because of its higher proportion. House damage by elephants too has been a very sensitive issue.

While a large proportion of the respondents opined that compensation was usually not quite enough, what they found more difficult was that it took rather long for the assessment of loss to be made and even after it had been done, payment was sometimes delayed in the process of sanction and sometimes because of inadequate availability of funds. Sustaining their families in the immediate aftermath of crop damage or after the elephants had broken into their houses and consumed the paddy they had stored was often difficult, particularly when they had nothing to fall back upon during the time lag from loss of crop till payment of compensation. And if it was winter when elephant raids increased because of the harvesting season or the rainy season when the skies opened up, the affected families had to face the wrath of nature as much as the fury of the marauding elephants. Of these respondents, 62 per cent felt that the local wildlife authorities could be more empathetic and sensitive to their difficulties. Interestingly, 4 per cent of the respondents actually opined that retaliatory killing of elephants was justified because they found themselves being driven to the edge.

Of the wildlife officials who responded to the questionnaire, nearly half felt that the compensation paid for damage was adequate though payment did sometimes get delayed through official processes. The average time taken from incident to payment was generally around six months but sometimes took over a year and more. On the crucial matter of coordination between the relevant enforcement agencies, 46 per cent or just under half felt it was proper. Regarding the query whether field officials had the authority to initiate innovative steps on their own, only 12 per cent responded in the affirmative.

The questionnaire for police officials revealed as much from what was not said as from what was, and the ostensible reason appearing to be that police officials did not deem such conflicts to be a matter that was overly their responsibility. In the background of law and order situations arising after elephant attacks which the police were required to handle, 60 per cent policemen thought coordination between them and wildlife officials was not quite adequate while 55 per cent opined that they had their own problems to worry about and felt burdened by the frequent law and order fallout of what was essentially the responsibility of the wildlife authorities.

Two broad inferences that emerged from the responses were that, on the one hand, there could be better coordination between the stakeholders for effectiveness in the field and, on the other, that there needed to be better clarity of job definition and role function.

Shown below is a hoarding at a blind curve on a stretch of highway in Kendujhar, among the more conflict-affected districts in Odisha because of extensive mining operations. Motorists are cautioned of this being the pathway of wild elephants and requested to inform the local forest range office of their movement if seen.

The picture shows an African elephant, *Loxodonta africana*, a species not found in India and different from the Asian variety, *Elephas maximus indicus*, which inhabits these forests. Domain knowledge and clarity of role definition is important for the credibility required to effectively resolve human–elephant conflict in the field.

Plate 23 Hoarding cautioning motorists of elephant passageway

In a positive step towards mitigating the frequent law and order situations caused by people after death from elephant attacks, however, the state wildlife authorities have initiated a welcome development. Eight forest divisions have been classified as high elephant depredation prone, eleven as medium depredation prone, and the rest as low prone. Five lakhs is proposed to be placed as revolving fund with each of the concerned divisional heads of high-prone zones, three lakhs with medium ones, and one lakh with the rest. Of the compensatory payment, 10 per cent is due is to be released to the next of kin within 24 hours on reasonable proof of identity of the deceased, and the next of kin, 25 per cent within 10 days on confirmation of procedural matters about the family affected, and the remaining 65 per cent on receipt of post-mortem and associated reports from the police.[10]

Even while this may not solve the problem entirely, it is a move forward to bridge the gap between the authorities and affected people.

Notes

1. *Daily Sambad Odia*, Bhubaneswar, 6 July 2008.
2. *Times of India*, Bhubaneswar, 9 December 2012.
3. 'The Delinquents: A Spate of Rhino Killings', *CBS News*, 11 February 2009, available at http://www.cbsnews.com/2100-500164_162-226894.html.
4. B. Behera, *Call of the Wild: Sanctuary Asia* 33 no. 3 (2013): 46–49.
5. N. Palei, B. Rath, and C.S. Kar, 'Population Status, Conservation and Management of Asian Elephants (Elephas Maximus) in Elephant Reserves of Odisha, India', *World Journal of Zoology* 8, no. 3 (2014): 292–98; C.K. Sar and S. Varma. 'Perspective Plan for Elephant Reserves of Odisha', *ANCF*, available at www.asiannature.org.
6. Harsha B. Udgata, 'Man-Wild Animal Conflict in Odisha', available at http://orissa.gov.in/e-magazine/Orissareview/2011/Feb-Mar/engpdf/60-64.pdf.
7. Available at http://www.orissadiary.com/CurrentNews.asp?id=14459; *The New Indian Express*, '14 Jumbo Corridors Identified to Reduce Man-Animal Conflict', Bhubaneswar, 28 August 2013.
8. *Express News Service*, 'Centre Seeks State Plans on Jumbo Corridors', Bhubaneswar, 2 September 2017.
9. Letter no.3647/4WL(G)4/2015 and Annexures dated 7 May 2016, Office of Principal Chief Conservator of Forests (Wildlife) & Chief Wildlife Warden, Bhubaneswar, Orissa.
10. *The New India Express*, 'Advance Payment for Human Kills by Jumbos', Bhubaneswar, 16 September 2017.

3

Elephant under Siege

The study of elephant deaths can be seen as a circle within a wider concentric circle and to reach the centre, one has to start from the circumference of the outer. The study of elephant deaths therefore begins with a study of the conflict between elephants and humans.

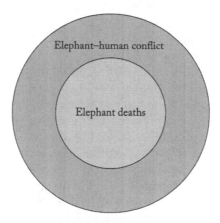

Figure 2 Context of study of elephant deaths

Nature has always had its own process of natural selection and survival of the fittest, both between members of a particular species and between the species and others—intra- and inter-species survival. Changes and adaptations are established by nature and become necessary in the ever-evolving mechanism of life, and are, therefore, made by such species for their survival and well-being. But with man's dominance and his ravenous intervention in nature's processes, the fragile balance that is determined by nature and is required to be maintained, gets disturbed. As in many other species, this is visible in the case of elephants.

The call for an international ban on ivory trade notwithstanding demand for and illegal trade in the commodity continues unabated, possibly more because of the restrictions. Both the *Loxodonta africana* bush elephant and *Loxodonta cyclotis* forest elephant of Africa continue to be targeted as they are the Indian *Elephas maximus indicus*, the last of the Asian variety left with ivory value. Despite there still being an estimated 4,00,000 or more elephants in Africa, fears of the forest elephant losing out in its bid for survival abound. In comparison, Indian elephants are around 27,000, but despite their obvious vulnerability in terms of numbers, there is a difference in the effect of poaching on the two kinds as far as consequences are concerned.

Both males and females of the African species have tusks, so when they are poached for ivory, their numbers dwindle but the sex ratio between males and females does not alter too much. In case of the Asian elephant, largest numbers of which are in India, it is only the male that has tusks. When tusk-bearing males are singled out for elimination, their numbers fall which not only drastically alters the gender equation that was so carefully determined by nature but also gradually depletes the gene pool of such tusked males for passing on to future generations.

Even while there are still bulls with big tusks, many of their kind have continued to be killed over the years and not many of those that remain are found with that kind of ivory. As the genes of the larger variety no longer being adequately around to be passed down to progeny, it is the lesser endowed male that remains to pass on his own. With even such males continuing to be singled out and killed, it is possible in the foreseeable future that their numbers too will fall. A possible situation may, therefore, not be very far away in the future when there would be no tusked males in the wild to breed and only tuskless males will remain to do so. The prospect of future generations of male Asian Indian elephants in the wild no longer having tusks is, therefore, a frightfully real possibility as has already happened with Asian Sri Lankan elephants.

As already mentioned, elephant deaths in India can be broadly categorised under the parameters of deliberate killing, accidental deaths, death in natural course, and from reasons not known. As the socio-environmental milieu in Odisha is representative of that in many large states of India where the twin problems of poaching and militancy exist, a study of the situation in the state would be largely reflective of that in such other states and the country overall.

Poaching has traditionally been done with guns, but use of poisoned arrows shot from bows has also been increasingly resorted to in order to escape the attention that the sound of the gun invariably attracts. In recent times, two other forms of killing have also been increasingly resorted to, both because they are soundless in operation and also because they cannot easily be traced back. One is the use of poisoned eatables deliberately left along the route known to be taken by elephants, or lacing with poison of areas with mineral-rich soil that elephants frequent. The other is deliberate electrocution engineered by laying out indiscernible live electric lines in their paths of movement. Since there is usually a time lag between lay out and execution in such instances, it is much more difficult to identify the perpetrators. Unlike poaching with guns or poisoned arrows where tuskers are invariably the targets, killing through poisoned eatables or deliberate electrocution often envisages cow elephants and juveniles falling prey because there is no guarantee of who comes across the eatables or in touch with the live wires first.

CAG Report

The office of the Comptroller and Auditor General (CAG) of India, after conducting a study in 2008, expressed concern at the rising rate of elephant deaths in Odisha. This report which was tabled on the floor of the State Legislative Assembly that year indicated that elephant deaths in the state had been averaging 32 per year between 1990 and 2003 but had risen to 56 a year between 2003 and 2008. Although the elephant population had risen from 1,841 in 2002 to 1,862 in 2007, 280 elephants had died during the period in the state that is home to 74 per cent of elephants in central India and 10 per cent of the country's tusker population, the CAG report said.[1]

The situation does not seem to have improved much. Actually, deaths seem to have only increased since then if official figures are to be believed. In his statement made on the floor of the Odisha Legislative Assembly on 27 August 2013, the Minister for Forest and Environment in the Odisha government said that from the beginning of 2000 until 2012–13, there had been 718 elephant deaths in Odisha.[2] Records maintained in the office of the Chief Wildlife Warden, Odisha, indicate that 529 elephants have died since 2008.[3] From 32 a year between 1990 and 2002 to 56 between 2003 and 2008, the rise to an average of 66 elephant deaths a year since then is consistent with the similarly increasing trend seen in respect of humans killed by elephants over the same period, referred to in the earlier chapter.

Following a query made about elephant mortality in the country, the Project Elephant Division of the Ministry of Environment, Forest and Climate Change, Government of India intimated statistics of total elephant mortality and deaths due to various reasons since 2010–11. As per the particulars received, Odisha has had the largest number of elephant deaths as also specifically of deaths from poaching in the past five years.[4] Karnataka, Assam, and West Bengal came thereafter.

Elephant Deaths since 2000–01

There have been a total of 896 elephant deaths in the state since the beginning of this century as per figures sought from and provided by the office of the Chief Wildlife Warden, Odisha.[5] Even while there might be scepticism regarding the accuracy of the numbers given the fact that not all deaths come to notice, particularly when they take place in deep and unfrequented jungle or where carcasses are clandestinely disposed of, it can at least be surmised that actual deaths would not be less than the figures projected because available figures would be based on elephants actually found dead. In the absence of authenticated statistics to the contrary, despite suspicion that there could be deaths happening that have not come to official notice, the figures officially provided will, therefore, be treated as the reference benchmark for the purpose of this study.

Going by the figures of 2015–16, a year after one and a half decades of the century, there would be reason to believe that elephants continue to die in increasingly greater numbers as the years go by. Seventy-six deaths were reported during the year as compared to the average of 66 a year for the preceding 8 years and 55 over the earlier 15 years. However, for a comparative study of trends, it would be desirable to have uniform time periods as base, and it is, therefore, proposed to divide the 15 years into three periods of 5 years each. Reference will be made to the 16th year as and when required.

In the 15 years from 2000–01 to 2014–15, there have officially been 820 elephant deaths in Odisha caused by the following 9 reasons.

Poaching—139
Poisoning—17
Accidents—64
Train hits—17
Accidental electrocution—69
Deliberate electrocution—59
Reasons not known—128
Natural causes—123
Disease—187

For a visual appreciation of the comparative death figures, the bar graph in Figure 3 will be illustrative.

It is seen that the highest number of deaths has been from disease, followed by poaching. The next highest figure is death from reasons not known while natural causes make up the next. Thereafter, it is accidental electrocution followed by accident, deliberate electrocution, and finally poisoning and train hits, and the last two figures being identical.

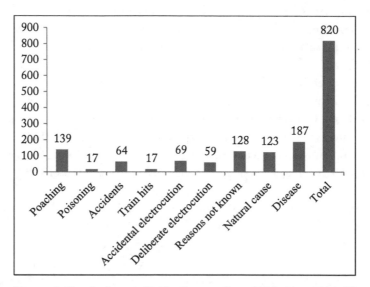

Figure 3 Deaths from individual causes from 2000–01 to 2014–15

In terms of the percentage proportion of each, the particulars may be illustrated by way of a pie chart as below.

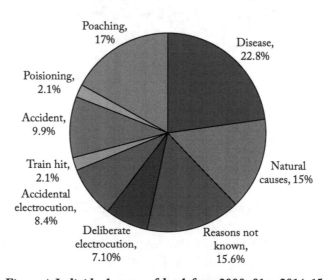

Figure 4 Individual causes of death from 2000–01 to 2014–15

Disease has taken nearly 23 per cent of all elephant lives while poaching has been responsible for 17 per cent. Unknown reasons make up a fairly high percentage, close to 16 per cent while natural causes make up 15 per cent. Poisoning that has been resorted to as another form of planned killing makes up 2.1 per cent, and deliberate electrocution, yet another form of premeditated killing of tuskers and crop raiders, 7.1 per cent. Accidental electrocution makes up a fairly large proportion of 8.4 per cent of the deaths, other accidents comprising nearly 10 per cent, and train hits 2.1 per cent.

For a clearer understanding of the broad causes of elephant deaths, it would be desirable to classify them under the four broad categories of deliberate killing, accidental deaths, death in natural course, and death for reasons not known. Deliberate killing would include poaching, poisoning, and deliberate electrocution while accidental deaths would refer to accidents and deaths from infighting, train hits, and accidental electrocution. Natural course would imply death from disease and natural causes, and reasons not known would mean cases where no definite opinion has been arrived at as to the cause.

A bar graph indicating number of deaths under each of these broad parameters is as below:

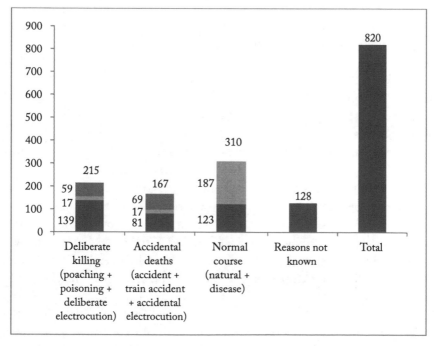

Figure 5 Deaths under four broad categories from 2000–01 to 2014–15

Even while the figure under each broad head is self-explanatory, the percentage composition of each as shown in the pie chart hereafter is illustrative.

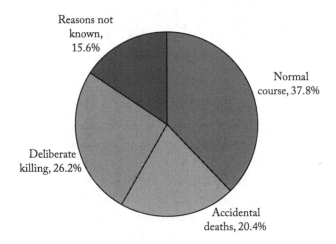

Figure 6 Percentage categories of death from 2000–01 to 2014–15

As may be seen, death in normal course comprises the highest proportion of close to 38 per cent. While deliberate killings make up over 26 per cent of all elephant deaths, accidental deaths comprise over 20 per cent. A grey area is death from unknown reasons which makes up close to 16 per cent. Even while it is a fact that delayed discovery of corpses, often in the form of skeletal remains or decomposed bodies, makes it difficult to identify the cause of death, the reason why such carcasses are discovered late tends to be a subject of some debate itself.

It is reasonable to assume that for each individual case, 15.6 per cent deaths from unknown reasons belongs to one of the three other categories—namely death by deliberate killing, from accident, or in normal course.

Going by the theory of probability in the absence of specific break-up of data in this regard, the 15.6 per cent deaths from unknown reasons would comprise deaths by deliberate killing, by accident, and in normal course proportionate to the individual percentage of each category in the overall death figure. Deliberate killing at 26.2 per cent of the overall death figure translates into 26.2/84.4 × 100 = 31 per cent of deaths where the cause is known, and the figure 84.4 arrived at by subtracting the unknown death percentage from the total, that is, 100–15.6. In other words, close to a third of elephants die from poaching and premeditated killing. Considering the fact that most of those killed are tuskers—only some of those poisoned and deliberately electrocuted happen to be unintended female or juvenile victims—the rate at which they are being eliminated is grim warning of the threat that they are in.

With deaths in normal course constituting 37.8 per cent of all identified deaths, they comprise in all probability $37.8/84.4 \times 100 = 45$ per cent of all deaths. As 22.8 per cent of identified deaths were from disease, it may be inferred that this form of death constitutes $22.8/84.4 \times 100 = 27$ per cent of all deaths by the same yardstick. One may, therefore, infer that over a quarter of all elephant deaths has been various diseases alone which suggests that the pachyderms here are not very healthy animals.

An examination of why the elephants are unhealthy throws up some interesting questions. In Kautilya's *Arthashastra* written around the second– third century BC during the reign of Chandragupta Maurya, the Mauryan empire which covered just about the whole of India as we know it today is said to have had nine 'Gaja vanas' or elephant forests. Elephants that inhabited the Indus basin and Saurashtra areas were of poor quality which probably explains why they disappeared in due course. The quality of elephants deteriorated from east to west, those found in central India being the best variety and the most useful for warfare. Significantly, this largest elephant forest in central India where the best elephants were found was the 'Kalinga vana' named after Kalinga, ancient name for Odisha. The trend continued over the centuries and during the Mughal times as well, the best war elephants being procured from these areas.[6]

Today, with Odisha having over 60 per cent of the elephants of central India, one wonders what would be the reasons that have caused such a decline in their standards of health to the extent that a third of them die from diseases. Even while the absence of good quality fodder caused by increasing denudation of forests would be a reason for elephants having a weaker immunity system, the periodical surfacing of diseases like anthrax amongst elephants and people living in the vicinity of forests alike is also an indicator that the disease is not entirely being wiped out.

Anthrax spores can remain latent for long period of time and resurface much later, years and sometimes even decades afterwards. There have been instances when herbivores that have eaten grass or leaves of plants that have grown where an anthrax victim died or was buried long years before have been infected by the bacteria. A possible reason for anthrax deaths to keep happening from time to time could be that animals dying of anthrax are not always identified and their carcasses burned, allowing the spores to remain dormant and resurfacing from time to time when conducive. That this could be a major killer among elephants is a distinct possibility even though it might not have reached epidemic proportions anytime for it to be declared as such.

The two major causes of elephant death, deliberate killing and disease, have claimed 35.8 + 31.2 = 67 per cent or two-thirds of all elephant deaths in the state of Odisha this century. Pending a more detailed analysis later in the chapter, a period-wise break-up of death figures would be useful.

First Five-year Period, 2000–01 to 2004–05

To understand the trends and nuances of the various forms of elephant death this century, it would be desirable to break up the first 15 years of the year into three periods of five years each. As phases and trends were not being entirely consistent over the 15 years, developments from one phase to another would make their understanding more inclusive. Taking the statistics of each block period separately, elephant deaths during the first five-year period from 2000–01 to 2004–05 under various categories could be depicted as under.

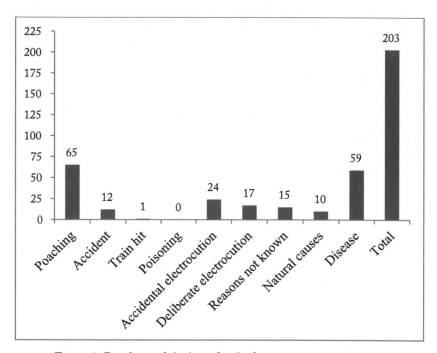

Figure 7 Break-up of elephant deaths from 2000–01 to 2004–05

Of the 203 elephant deaths on record in the first 5 years, the 65 cases of poaching make up the single highest cause. This works out to a rather high percentage of 32 per cent of the total deaths. Disease forms the next highest with 59 deaths at 29.1 per cent. A diagrammatic depiction of the percentage composition of different forms of death during the period:

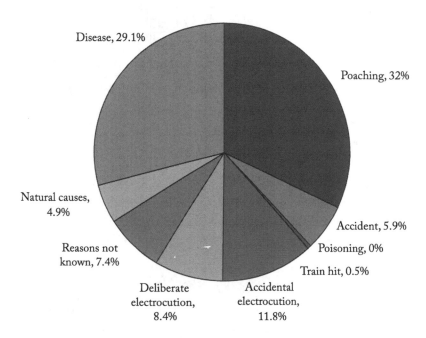

Figure 8 Percentage deaths from 2000–01 to 2004–05

Poaching at 32 per cent of the overall deaths reported is a high component, and together with deliberate electrocution at 8.4 per cent deliberate killings make up 40.4 per cent of all deaths. Death by poisoning has not begun as a mode of intentional killing as yet.

Death in natural course makes up 34 per cent of which death from disease is 29.1 per cent and from natural causes, 4.9 per cent. There has been just one case of death by train hit at 0.5 per cent of the total figure. Cases where the cause of death is not known make up only 7.4 per cent.

A diagrammatic representation of the proportion of deaths under the four broad categories of deliberate killing, accidental deaths, death in natural course, and those for unknown reasons would be as under.

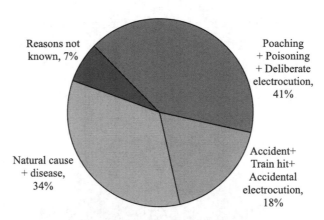

Figure 9 Broad categories of death from 2000–01 to 2004–05

What is particularly discernible from the above diagram is the very high percentage of cases of deliberate killing in the total number of deaths. Death in normal course also makes up a high proportion of which disease forms the bulk. The percentage of death from unknown reasons is not very high.

From the first half decade from 2000–01 to 2004–05 on to the next half decade from 2005–06 to 2009–10, some changing trends are seen.

Second Five-year Period, 2005–06 to 2009–10

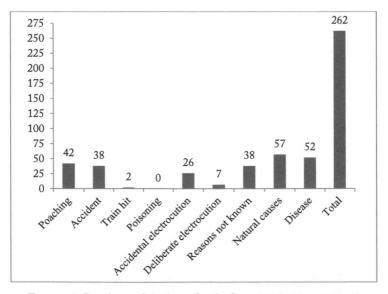

Figure 10 Break-up of elephant deaths from 2005–06 to 2009–10

Perusal of the bar diagram shows that there has been a significant jump in the total number of deaths from 203 in the first half decade to 262 in the second. The increase of 59 deaths in the overall number represents a substantial rise of 29.2 per cent from the earlier period.

Interestingly, death by poaching, the highest contributor in the first half decade, has dropped from 65 cases to 42, a fall of 23 cases and a percentage drop of 35 per cent. Cases of deliberate electrocution have also come down from 17 to 7, and with no case of poisoning having been reported this half decade as well, the number of deliberate killings which was 82 in the first period (poaching 65 + deliberate electrocution 17) has come down to 49 (poaching 42 + deliberate electrocution 7). This drop by 33 cases represents a huge fall of 40.2 per cent in deliberate killing of elephants implying thereby that preventive measures could have borne results or, alternatively, that they could possibly have been reflected under some other category given that the overall death figure has increased considerably. In this regard, figures under death for reasons not known draw attention. From 15 cases in the first half decade, such deaths have increased to 38 in the second, an increase of 23 deaths representing an enormous rise of 153.3 per cent.

Pie charts of the percentage break-up of individual causes of death and of the four broad categories given below would show the changes that have come about in the percentage composition from periods I to II.

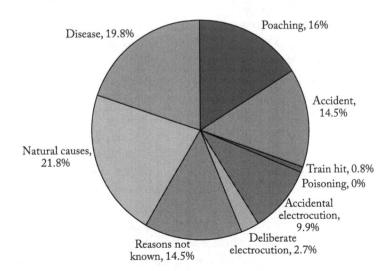

Figure 11 Percentage of individual causes of deaths from 2005–06 to 2009–10

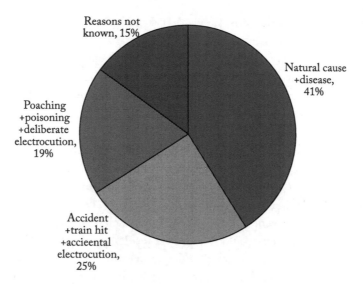

Reasons not known, 15%

Natural cause +disease, 41%

Poaching +poisoning +deliberate electrocution, 19%

Accident +train hit +accieental electrocution, 25%

Figure 12   Category break-up of deaths from 2005–06 to 2009–10

When one compares the percentage-wise representation, the percentage of poaching deaths is seen to have decreased from 32 to 16 per cent and of deliberate electrocution from 8.4 to 2.7 per cent. Corresponding to these particulars, the overall percentage of deliberate killings as seen for periods I and II has come down from 41 per cent to just 19 per cent. This would prima facie suggest that anti-poaching measures and steps to curb deliberate killing of elephants have been successful. However, it would be interesting to examine if fluctuation in any of the other categories could be relevant in this context.

Significantly, the proportion of cases where death has occurred for reasons not known has almost doubled from 7.4 per cent in period I to 14.8 per cent in period II, and in terms of numbers, it has more than doubled from 15 to 38.

Why the number of deaths from unknown reasons should have suddenly risen so sharply could be explained by the fact earlier mentioned that carcasses have possibly been detected too late for the cause of death to be identified. Incidentally, this was the period when there was heightened militant activity in the state, armed left-wing extremists operating almost entirely from forests. With the increased threat perception from militants in such forests, wildlife officials were understandably reluctant to enter them. Consequently, elephants that died in these forests were discovered far too late for an opinion to be formed about the cause of their deaths. This probably explains why, when the total number of deaths rose from period I to period II, poaching cases dropped sharply even as death from unknown reasons rose. Even while anti-poaching measures might have had their own contribution in bringing down such cases, the substantial increase in the total number of deaths creates suspicion that a sizeable portion of the increased cases of death from unknown reasons could actually have been instances of deliberate killing discovered at such belated stages that the cause of death could not be conclusively established.

Third Five-year Period, 2010–11 to 2014–15

Moving on to the third period in the overall time frame of this study, the break-up of elephant deaths from 2010–11 to 2014–15 reveals consistency of changes in major forms and the inferences that had been drawn from the earlier trends.

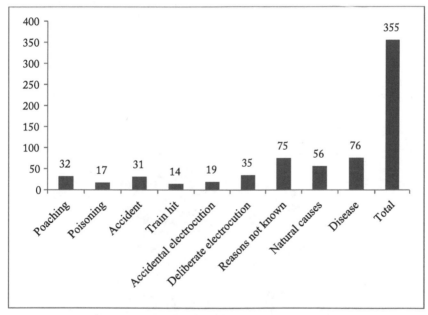

Figure 13 Bar graph of deaths from 2010–11 to 2014–15

From 203 deaths in period I to 262 in period II, the number of elephant deaths in period III rose to 355. This means that the total number of elephant deaths in the state rose by 29.1 per cent from the first half decade of the century to the second, and by 35.5 per cent during the half decade thereafter. This was irrespective of the fluctuations in the individual causes elephants supposedly died from.

Percentage proportions of individual causes and the broad categories in the total number of deaths in period III are illustrated in the following pie charts.

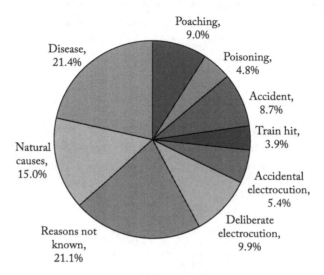

Figure 14 Percentage of individual causes of deaths from 2010–11 to 2014–15

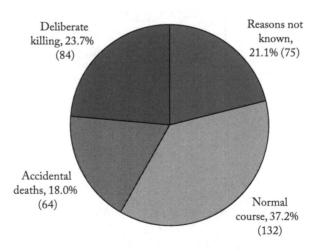

Figure 15 Broad causes of death from 2010–11 to 2014–15

From a comparison of figures of deliberate killings which are essentially preventable deaths, it is found that poaching which constituted 32 per cent of the total deaths in period I fell to 16 per cent in period II and further down to 9 per cent in period III. Deliberate killing or preventable deaths as a whole constituted 41 per cent in period I, came down to 19 per cent in period II, and rose a bit to 23.7 per cent in period III. Though poaching figures have per se fallen, poisoning cases have risen from 0 per cent in the first two periods to 4.8 per cent in the third. Deliberate electrocution was 8.4 per cent in period I, fell to 2.7 per cent in period II, but rose to 9.9 per cent in period III.

For a comparative picture of the three periods, the following bar diagram is illustrative.

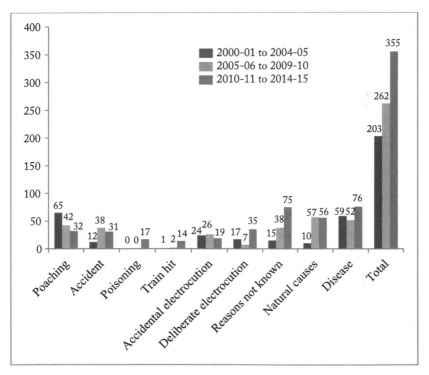

Figure 16 Comparative bar graph for the three periods

Individual heads that draw attention are the three under deliberate killing, namely poaching, for the sharp decline it has shown over the years, as also poisoning and deliberate electrocution which have shot up in a big way during the third period. The latter two also explain why, despite the gradually decreasing figures of poaching, the percentage of deliberate killing has risen from 19 to 23.7 per cent from period II to period III. Even while poisoning and deliberate electrocution seem to be increasingly resorted to as a new form of deliberate killing, another head that has shown enormous increase and is suspected to be linked to poaching figures is death from reasons not known.

Interestingly, elephant death figures for the year 2015–16, on completion of the third period ending 2014–15, confirm the inferences that one draws from the comparative study of the three five-year periods. The average number of elephants that died each year in the period between 2000–01 and 2004–05 was 41 which rose to 52 in the second period from 2005–06 to 2009–10. In the third, from 2010–11 to 2014–15, the average number of deaths per year further rose to 71. There were 76 elephant deaths in 2015–16 which are consistent with the increasing number of deaths each year over the years.

Significantly, the average number of elephants officially shown to have been killed by poachers was 13 every year in the first period, fell to 8.4 in the second period, and further to 6.4 in the third. In 2015–16, the number of elephants killed in poaching is seen to be just 2. Conversely, the number of elephants that died from unknown reasons averaged three each year in the period from 2000–01 to 2004–05 and went up to nearly 8 in the second period from 2005–06 to 2009–10. In the third period from 2010–11 to 2014–15, it went further up to an average of 15 a year and significantly has gone up further to 21 during the year 2015–16, revealing a similar upward consistency as poaching figures have showed downward. That the two trends could possibly be linked will be elaborated further in course of this chapter.

An inference that can be drawn from the decrease in poaching cases but increase in deliberate killing from the second to the third period is that instead of killing directly with guns, poachers have resorted to indirect methods like electrocution and poisoning. Even while use of poisoned arrows continues to be resorted to as a form of poaching to avoid detection from the sound that gunfire attracts, killing with poisoned arrows is not generally an indigenous trait and is mostly linked to poachers hired from outside the state by ivory traffickers. Certain tribes of forest dwelling people in the north-eastern part of the country have knowledge of extracting poisonous concoctions from some local roots and herbs, and, over the years, many of them have been hired for such work.

Some recent incidents in Kenya where elephants were being machine-gunned for ivory would suggest that the use of poisoned arrows is not necessarily a local phenomenon or a dated practice. Two of the Africa's most iconic and widely photographed elephants, 'Torn Ear' who was one of the few left in the continent with tusks weighing over 100 pounds[7] and 'Satao' who had tusks over 2 metres long and belonged to a breed genetically having unusually large tusks,[8] were both killed in 2014 by ivory poachers with poisoned arrows.

Leaving poisoned eatables along the pathway of elephants has been resorted to as an alternative method of deliberate killing in recent times. There was no case of poisoning over 10 years of the first two periods, but it appeared as a sudden burst of 18 cases in the third from 2010–11 to 2014–15. Of these, 16 were in the first 3 years with 9, 4, and 3 cases, respectively, while there was one each in the remaining two. There has been no case of poisoning in the year 2015–16, which suggests that after a sudden spurt when this method of killing was tried out and experimented with, it may have gradually lost favour.

Apparently, the difficulty in tracing the crime back to the perpetrator was the advantage that this method was believed to have been provided to the poacher. The time lag between laying out the bait and the consequent death of the elephant allowed the poacher to stay in the background and come for the tusks only when the coast was clear. If the killing was detected, he could simply remain in the background and avoid detection. However, which animal got to the poisoned eatables first was a matter of chance. The relatively low probability of actually bagging tuskers and the unwanted large number of female and juvenile elephant casualties apparently did not make the effort very worthwhile, particularly with the attention that this new form of killing drew.

Likewise, having laid live electric wire along elephant routes to deliberately electrocute them, electricity distribution networks now extending through dense forest areas as well, poachers could move in later after the elephant had been killed. The time between layout and death in these two forms allows the poachers to remain behind the scene during the actual killing which makes detection difficult and the likelihood of the killing being traced back to them in criminal investigations equally so. Perusal of statistics shows that there were 17 cases in the first period, 7 in the second, and as many as 35 in the third. In other words, the average per year was slightly over three in the first, a little over one in the second, and suddenly as high as 7 in the third period.

In the year 2015–16 too, there have been 9 cases of deliberate electrocution which suggests that the increasing trend of recent times is continuing. But the handicap that applies to poisoning is relevant here as well because tuskers are not necessarily the ones killed as female and juvenile elephants often become unintended victims of collateral killing when they come in contact with the laid-out live wire before the targeted tuskers do. Even while live electric wire is often laid out by poachers along routes, solitary tusker elephants are known to take and such efforts do indeed result in their death sometimes. Deaths under this column have been rising more because farmers have been increasingly resorting to this tactic to keep animals off their crops. Such killings, though deliberate, are therefore more of preventive killings.

With regard to deliberate killing of elephants, there were a total of 154 criminal cases registered until 2015–16 in course of which 417 people were arrested. With 226 elephants having been deliberately killed from 2000 to 2001 until date as per official records, 141 to poaching, 68 to deliberate electrocution, and 17 to poisoning, perhaps all such deaths have not led to registration of criminal cases. Alternatively, there might be instances when one case might have been registered for more than one animal killed.

The particulars of criminal cases and corresponding arrests until 2015–16 may be represented graphically as below. A feature that tends to draw attention is that there has been no particular consistency between cases registered and accused arrested over the years. 15 cases had reportedly been handed over to the State Crime Branch for investigation given the sensational nature and sensitivity of some. From a perusal of the graph below, it is seen that the number of cases registered was highest at 43 in 2011–12, fell to 23 the next year, further down to 15 the next, and to 7 and 8, respectively, in the two years thereafter. Even while the number of people arrested had increased, the decreasing number of cases detected as deliberate killing does tend to raise some degree of scepticism.

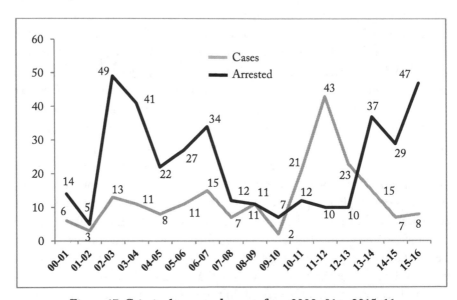

Figure 17 Criminal cases and arrests from 2000–01 to 2015–16

Organised Poaching

In continuation of the deliberations from this chapter on the possible connection between the diminishing instances of poaching and the corresponding increase in deaths from unknown reasons, statistics are revealing.

In comparison with poaching deaths that decreased from 65 to 42 to 30 in the three five-year periods, deaths from unknown reasons rose from 15 to 38 to 80. While the average deaths from poaching fell from 13 per year in the first period to a little over 8 in the second, and 6 in the third, the average of unknown deaths per year rose from 3 in the first period to nearly 8 in the second, and 16 in the third. In the year 2015–16 following the completion of the third period, cases of recorded poaching fell to a mere two while those from unknown reasons rose to 21. Viewed in terms of percentage, the proportion of poaching deaths decreased from 32 to 16 to 7.8 per cent over the three periods while deaths from unknown reasons increased from 7 to 15 to 20.6 per cent, their respective positions having inversed.

It might be rather simplistic to jump to the conclusion that anti-measures have succeeded in bringing poaching figures down. In the backdrop of the increase in overall number of deaths over the years, the sharp increase in cases of death from reasons unknown calls for some examination. Ordinarily, it is not very difficult to ascertain the cause of death when the elephant is found soon after it dies and body injuries or other symptoms, as are physically available, can be detected. If physical examination does not reveal much, a post-mortem examination can certainly do. However, if the carcass remains undetected for a relatively longer period, there is time for scavengers like wild boar that proliferate in most forests to strip the flesh of the body and leave no trace of injuries or symptoms the elephant might have had on it. When only skeletal remains are left, it is difficult to know what the animal died of.

It can, therefore, be inferred with a reasonable degree of certainty that instances where the cause of death remains unknown happen primarily, if not entirely, when the death is detected at a belated stage. This usually happens when elephants die in areas which are not frequently accessed by people, and their carcasses are consequently discovered long after they die. In other words, the progressive rise in the number of deaths from unknown reasons suggests that more and more of elephant deaths are being detected very late. This would suggest, in turn, that they are dying in areas that are not visited much which in turn would mean that this is happening in inaccessible forest areas.

If there indeed is a correlation between the systematic drop in the number of poaching deaths and the corresponding increase in deaths from unknown reasons as seems evident, it is a matter of concern because it suggests that elephant killers have become wiser and are able to cover up their tracks long enough for poaching cases to be passed off as death from unknown reasons. Coming in the wake of the likelihood that poisoning and electrocution have been resorted to sometimes as alternatives to direct poaching in order to avoid detection, it is important that this subterfuge does not lead to the complacency of presuming that poaching has been controlled. But if the increase in deaths from unknown reasons is because carcasses are discovered in forests belatedly, the reason why forest and wildlife officials should not be able to detect them in such forests more promptly calls for some scrutiny.

A former official of the state wildlife organisation, while analysing the increasing number of elephant deaths, had mentioned what he believed to be of relevance. It is generally known that elephants use specific routes for movement, and even when scarcity of food caused by habitat degradation forces them to move elsewhere, they return to the places they had started from. This predictability of movement helps poachers who generally use guns for killing but have also been using poisoned arrows and have now been baiting them with poisoned eatables. Information of four dead elephants lying in the core area of the protected Simlipal Tiger Reserve was not given by officials but by an NGO that specifically looked for them after getting intelligence inputs. Had the NGO not provided the information, the carcasses would perhaps have been detected far too late for knowing that they had been poisoned. The official inability to detect such cases could suggest that this could be happening in other forests as well. Preliminary investigation into the poaching of three elephants for tusks in Mayurbhanj and of a tusker in the border area of Kalahandi and Rayagada suggested the presence of at least four major poaching gangs in the state, and one of which had engaged a Namdhari tribal from the north-east. While the said tribal poacher was caught, others engaged by the other gangs managed to give the authorities the slip when things became hot.[9]

Even though the presence of professional poachers of the Lishu Tribe of Arunachal Pradesh and Manipur has been noticed in Odisha since 1994, it was only after 2007 that there has been confirmation of their continued visits to various parts of the state for killing tuskers. During the investigation of the tusker killing in the border jungles of Kalahandi and Rayagada in 2007, the apprehension of a tribal involved in the killing and removal of tusks revealed, not only the frequent visits of such poachers from the north-east, but also their extensive coverage of various forests of the state. These poachers have traditionally used a poisonous concoction extracted from certain local jungle plants and used it through arrowheads shot into tuskers. Having shot several arrows into the elephant, they wait for the poison to take effect, and after it dies, they remove the tusks and pass them on to predetermined conduits who in turn carry them outside the state.

That this has been continuing unabated is evident from the fact that such poacher tribals from the north-east were seen in Mayurbhanj and Balasore districts in recent times, which was widely reported in the media. An extract from one of them, 'North-East Poachers sneak into Simlipal, Kuldiha Forests' published some time back is revealing.

Close on the heels of information that Lishu poachers hired from Arunachal Pradesh were responsible for the death of 2 tuskers in Simlipal Tiger Reserve sometime back and possibly two more in Kuldiha and Satkosia thereafter, and a group of 4 inter-state poachers from the north-east reportedly sneaked into Simlipal and Kuldiha forests around 10 October 2012. Reports of their presence in Baripada have been confirmed by the local wildlife authorities. Confirming suspicions that organised killing of elephants for their tusks continues to take a huge toll on the tusker population in the state, the group members recently spotted by locals are suspected to be Lishu tribals from Arunachal Pradesh or Manipur who are known to kill elephants with arrows poisoned with extract of certain roots and herbs. Two Lishu tribals from Manipur were arrested in 1994 for mass killing of elephants in Simlipal while at least 3 others who were with them are believed to have evaded arrest and escaped. A similar gang was apprehended at Bhawanipatna for killing of a tusker and removal of its tusks in 2007 which confirmed suspicions that not only are the activities of such poachers from the north-east widespread in Odisha but also that they have been operating consistently over the years. It is learnt that at least 45 elephants have been killed in the Simlipal region alone in the past 3 years.[10]

With reports suggesting the widespread presence of organised gangs indulging in illegal ivory trade in Odisha, the state government decided to hand over some sensitive cases having state-wide ramifications to the crime branch of the state police. Even though a total of just 15 cases were taken over by the crime branch over the last 16 years,[11] investigations have revealed that various organised gangs are active in various parts of the state. One such syndicate has been active in the Simlipal Tiger Reserve and adjoining areas like Kuldiha and Karanjia in the northern part of Odisha while another has been operating in the Baramba, Narsinghpur, Nayagarh, and Daspalla forests of the eastern part of the state.

There are also syndicates operating in Satkosia and in the Bhanjanagar–
Daringbadi areas in the central and southern parts of the state. Organised
local gangs spread over the entire state are believed to have been hiring the
services of Arunachal Pradesh Lishu tribals to silently hunt elephants with
poisoned arrows, extricate their tusks, and have them handed over to couriers
who transport them outside. Use of firearms continues to be resorted to despite
the obvious attention that their noise draws. The hired poachers are reported
to have stayed in hotels in the Mayurbhanj area on a few specific occasions,
removed the tusks after the killing, and sent them to Kolkata.

Even though investigation in some earlier cases had revealed that the gangs
in Odisha were conducting their ivory smuggling business through hired
poachers from outside the state, not much headway was made in apprehending
the killers because of alleged lack of inter-state coordination. It was reported
sometime back that 221 elephants had died for various reasons in the past four
years of which about 40 per cent of the deaths were suspected to have been for
tusks. Simlipal alone reported over 20 deaths drawing sharp reactions from
the central government and prompting investigations by the Wildlife Crime
Control Bureau. Information from the Ministry of Environment and Forests
suggests that the higher proportion of tuskers in Odisha makes it a favoured
destination for poachers, and as the tusks of Odisha elephants are hardier and
heavier even if smaller, they fetch more value.[12]

A perusal of the cases taken up for investigation by the crime branch of the
state police is interesting. Even while wildlife criminals have been arrested
and charge sheeted to stand trial in the court of law, it is found in just about
all cases that the weapons of offence seized have either been country-made
guns or bows and arrows.[13]

Under the circumstances, the question that begs to be asked is if there is
more to organised poaching than this. There were extensive reports in various
sections of the media and among wildlife activists sometime back that the state
has been seeing a steady decline in the tusker population. It was alleged that
at least a fourth of the elephant deaths each year was on account of poaching,
and the targets of such poaching invariably were tuskers. Importantly,
while the reports highlight the wide network that the poachers have in the
country, they also seem to confirm suspicions that some of them even operate
internationally.[14]

In the last week of July 2013, a poaching gang led by its mastermind
from Nepal reportedly shot dead a tusker in the Bhanjanagar–Sorada area of
Ganjam, and after taking away its tusks, they shot and injured another on 8
August 2013. Even while an accomplice from a local village was arrested and a

country-made gun and some ammunition were seized from him, he reportedly informed during interrogation that the kingpin from Nepal escaped along with two others.[15]

The operation is simple: kill one tusker, extract its tusks, and make arrangements for routing the ivory out of the state, and the job having been done, turn to the next one after the other. From killing to final disposal of the tusks taking a mere two weeks and the poachers targeting tuskers one after another, the high vulnerability of tuskers does not need elaboration.

A publication titled, 'Elephant deaths rampant in Orissa' highlighting that poaching for ivory is the major reason for high casualty of elephants, quotes senior officials of the State Wildlife Wing describing it as a three-stage operation where killings are done locally: the tusks sent to national-level operators outside the state and from there to different destinations in the world primarily via Nepal. A significant feature of this report based on media briefing by senior wildlife officials, and circulated through a *Press Trust of India* release is that such senior officials have stated officially on record that they are facing difficulties countering poaching because poachers are equipped with the latest weapons that are more sophisticated than the ones wildlife officials use.[16]

Poaching and Organised Crime

Available official records of cases indicate that the weapons used in the poaching cases investigated or inquired into by the wildlife authorities have largely been poisoned arrows or country-made guns. Review of the important syndicate poaching cases investigated by the crime branch of the state police too reveals that the most potent weapons that are found to have been used and seized are country-made guns. Given the obvious discrepancy between wildlife officials themselves admitting that they are not equipped to match the sophisticated weapons of poachers and the fact that only country guns are found to have been used even in the more important cases of poaching detected, it is possible that there is a more serious side to the problem that has not come to the forefront yet.

It is in the background of the confirmed presence of poaching syndicates in the state, the admission by wildlife officials that poachers are equipped with the latest weapons and have links with the illegal, international wildlife trade that the disturbing revelations of Boston based International Fund for Animal Welfare (IFAW)'s report of 2013 become relevant in the Odisha and Indian context.

Credible and knowledgeable international authorities now believe that the problem of poaching and illegal wildlife trade is no longer just an issue of wildlife and environment management but has come to cover the more important matter of state security. The IFAW in its report 'Criminal Nature— Global Security Implications of the Illegal Wildlife Trade' released on 24 June 2013 has highlighted the dangerous links that have reportedly now come to exist between poaching and organised crime. The report draws attention to elephants being machine-gunned in central African countries for ivory selling as high as $1,000 a pound and rhinos driven nearly to extinction for horns that cost more than gold and platinum. The illegal international trade in endangered species has integrated with organised crime and militant groups worldwide, the report has warned. In the past half decade, wildlife crime has reportedly grown into the fourth largest branch of illegal international trade behind only narcotics, counterfeiting, and human trafficking. Now worth $18 billion annually, the black market in animals and their parts, notably ivory and furs, threatens to eradicate many of the most iconic of species such as the elephants, rhinos, and tigers.

'Within the last few years, poaching has grown tremendously from one-off killings to wholesale massacres', IFAW's Beth Allgood has been quoted as saying, 'We can't see this as an environmental problem anymore when it has grown into a criminal and security one.'

The report comes as international observers have become more concerned about links between the illegal animal trade and terror groups in Africa and Asia. In November 2012, the US Secretary of State Hillary Clinton declared illegal wildlife trade a security threat. In May, UN secretary-general Ban Ki Moon released a report linking militant activities in central Africa to the illegal ivory trade and slaughter of elephants. Worldwide demand for illegal ivory has driven its price sky-high and made elephants even more vulnerable. Overall, the report draws a picture of militant groups worldwide increasingly turning to mass poaching to supply international organised wildlife smuggling rings in exchange for arms. Rhinos are being liquidated for their horns and elephants for their tusks, while a tiger is killed almost every day in India where just a few thousands are left in the wild. China, the US, and Europe are the leading markets for the illegal trade, states the report.[17]

It was around this time in October 2013, that the Los Angeles-based 'Elephant Action League' reportedly established after an elaborate undercover probe that the Somali terror group, Al-Shabaab, which took responsibility for 21 September 2013, Westgate Mall carnage at Nairobi, Kenya that killed

67 and injured 175 and used proceeds from the illegal ivory trade to fund its terrorist operations. The money trail connecting elephants massacred in Kenya was reportedly tracked to ivory buyers in Asia through this terror group, an offshoot of the now deceased Osama bin Laden's Al-Qaeda, whose returns from the trade are so high that it pays its terrorist mercenaries more than the Somali government pays regular army soldiers.[18]

It was on 2 April 2015, a year and a half after the Nairobi Westgate Mall Terrorist Attack that heavily armed gunmen attacked Kenya's Garissa University College and killed 148 and injured 79, most of them students. Al-Shabaab took responsibility for this bloodbath too.[19]

With countries like the Central African Republic, Chad, South Sudan, Democratic Republic of Congo, and Cameroon in the grip of intense poaching and bloody internal strife, armed militants are reportedly being funded by the ivory trade in a big way. The enormity of the problem can be gauged from the fact that the estimated 1,00,000 elephants are believed to have been killed by poachers in the last few years, and though the elephant casualty figure in 2015 had fallen to 20,000, Central Africa has reportedly lost 64 per cent of its elephants in a decade.[20] Most of the ivory finds its way to buyers in Asia and particularly China. Indeed, data released on UN World Wildlife Day in 2016 indicate that more elephants are being killed than are being born each year even though annual death figures have reportedly been coming down since 2015. Interpol believes that the wildlife trade has a turnover between $10 and $20 billion and is next only to drugs, human, and arms smuggling.[21] Also, as per figures of the International Union for Conservation of Nature (IUCN), rhinos too are being killed in increasing numbers for their horn over the past five years and a record of 1,338 died in 2015.[22]

IFAW's 'Criminal Nature' report probably is the first major public document on the subject. Less than a fortnight after its release, the official statement of India's Forest and Environment Minister at the valediction function of a seminar on 'Asian Big Cat Related Crimes' organised by Interpol and the Central Bureau of Investigation in New Delhi from 1–5 July 2013, brings added significance. Confirming evidence that some gangs involved in the illegal wildlife trade in India have links with terrorist organisations, the minister emphasised—before the professional body of wildlife crime investigators, officials, and activists—that such trade is a big business running into millions of dollars and is conducted the same way as the illegal drugs and arms trade are. Elephants in India continue to be hunted down for ivory, tigers for their skin and bones, and rhinos for their horns.[23]

Organised Poaching and Left-Wing militancy

Even while the scale of the illicit wildlife trade in Africa is perhaps a lot higher than what might be happening in India, at least in terms of numbers, the modalities of the trade here are a matter of conjecture as not much study appears to have been done on the situation in this country. Seen in the context of the 'Criminal Nature' investigation report that militant groups in Africa are arming themselves through the high return ivory trade, the possibility of a similar situation prevailing in India in the backdrop of the Indian Forest Minister's statement regarding links between terrorists and poaching gangs is not unreal. In this regard, the situation in the state of Odisha, which has been subject to the key elements of both armed militancy and poaching, would perhaps be a representative study of what could be happening elsewhere in the country too.

Odisha has been seeing an extreme form of violent militancy from left-wing extremists since the beginning of the century, the peak periods were from around 2003–04 to 2012–13. With the former Prime Minister of the country calling left-wing militancy the biggest threat to the security of the country, a recent study by the acclaimed New Delhi-based Institute of Defence Studies and Analyses (IDSA) has confirmed that Naxalites, who are spread over one-third of the India's land mass, are located mostly in forests.[24] Dense forests and wildlife sanctuaries are preferred by militant extremists all over the country for better cover and lesser chances of detection. It does not require special investigation to discover that poachers too operate in such forests and wildlife sanctuaries. This raises an important question.

In the backdrop of confirmed links between terrorist militancy and large-scale poaching in Africa, do poachers and militants found in and operating from the same forests in India operate independently or have links? More significantly, is there some common ground where militants could be turning to or hiring poachers as a source of funding or conversely poachers turning to militants for protection? This becomes particularly significant because wildlife officials have been reluctant to enter forests where militants have taken shelter. This probably also explains why such a large proportion of elephant deaths is attributed to reasons not known when many could well be cases of poaching.

Even while traditional poaching syndicates might be hiring Lishu tribals from the north-east to kill tusker elephants with poisoned arrows, the statement of the minister for Forest and Environment, Government of Odisha, in the Odisha Legislative Assembly that 57 of the 73 elephants killed by poachers from 2005 to 2011–12 have been shot dead is significant.[25] The fact that this was the period when armed left-wing extremist activity in Odisha was high, the revelation that poaching and militancy are closely linked, the report that militants make supplies to international wildlife trade rings in exchange for arms, the fact that left-wing militants in Odisha have sophisticated arms, and the admission by the state's wildlife authorities that poachers in Odisha also have the latest weapons may not all be coincidental.

The question whether there are links between poachers and left-wing extremists in Odisha or if the Naxalite militants are themselves resorting to poaching to fund their need for sophisticated weapons could perhaps be better appreciated through some map illustrations. While doing so, and in the context of the India's Forest and Environment Minister's statement that poaching gangs in India have links with terrorists, the statement of the former Home Minister of India that Naxalites are terrorists needs to be kept sight of.[26]

As per the reports of the state wildlife authorities, elephants are now seen in 28 of the 30 districts of the state.[27] From a stage when they were found in about 16, their spreading to all districts except Jagatsinghpur and Kendrapara is not so much because they have been increasing rapidly in numbers as for the fact that human intrusions in the districts they had earlier been confined to have forced them into areas they had not been getting into before. They are now found in close to all the districts, sometimes as full-time residents and sometimes temporarily. Their presence in various districts of the state can be illustrated in the map of Odisha as below.

Figure 18 Elephant-presence map in the districts of Odisha

Poaching has been found to be generally more in areas which elephants treat as their abode or those they habitually visit rather than areas that they may be passing through. From a perusal of poaching cases on record in various territorial and wildlife forest divisions from 2000–01 till date, the districts of Odisha where poaching exists can be illustrated as in the map shown next.

From official records in the office of Odisha's Chief Wildlife Warden 2000–01 onwards, it would appear that elephant poachers have been active in 18 of the 30 districts of the state, though fringe areas of some of the other districts cannot entirely be ruled out. These districts are Mayurbhanj, Keonjhar, Balangir, Cuttack, Kandhamal, Sundergarh, Khurdha, Deogarh, Dhenkanal, Ganjam, Sambalpur, Angul, Boudh, Kalahandi, Rayagada, Balasore, Gajapati, and Nayagarh.[28] Since there are districts other than these where poaching cases may not have been officially recorded but death from unknown reasons might actually have included instances of poaching, it may be inferred that these districts are the confirmed minimum number where poaching does exist.

Figure 19 Elephant-poaching map of Odisha

Coming to the presence of left-wing extremism in the state, 19 districts of Odisha have been declared by the Government of India to be Naxal-affected or eligible for security related expenditure status. These are those where left-wing militancy is deemed serious enough for the Government of India to bear the responsibility of funding all combat operations irrespective of which the state is.

These districts where armed militancy is high are Mayurbhanj, Keonjhar, Sundergarh, Sambalpur, Kandhamal (Phulbani), Nayagarh, Deogarh, Bargarh, Balangir, Dhenkanal, Jajpur, Ganjam, Kalahandi, Rayagada, Gajapati, Koraput, Nowrangpur, Malkangiri, and Nuapada.[29] Even while such left-wing militancy has abated in a few of these districts because of pressure from security forces, it has spread to new forested areas like Angul and Boudh districts where elephant poaching has existed but which had hitherto been free from armed militancy.

The map below illustrates the presence of left-wing extremism in Odisha's districts.

Figure 20 Maoist-militancy-affected districts of Odisha

When one tallies the 21 Naxalite-affected districts of Odisha with the 18 where poaching officially exists, 15 are found to be common to both. These are Mayurbhanj, Keonjhar, Balangir, Kandhamal, Sundergarh, Deogarh, Dhenkanal, Ganjam, Sambalpur, Kalahandi, Rayagada, Gajapati, Nayagarh, and Angul. Since Naxalites are known to be taking shelter and operating from forests, the presence of poachers in these forests is of significance. Conversely, poachers had already been operating in the forests of districts like Angul before militants moved into them.

Some of the districts where Naxalite presence is high are also the ones where there is high incidence of poaching, like Mayurbhanj, Keonjhar, Sundergarh, Dhenkanal, Sambalpur, Kalahandi, Gajapati, Deogarh, and in recent times, Angul and Boudh. This is an interesting situation because one would think that poachers would hesitate to enter forests where armed left-wing militants hold sway. If poaching gangs operate freely in these forests or conversely militants start operating from forests that had only seen poaching before, the inevitable inference is that there could be some tacit collusion between them.

The districts where Naxalites and poaching coexist are illustrated in the map below. A significant feature of the commonality of districts where there has been poaching and those that have seen militancy is that the developments have not sprung up together but happened over a period of time. Some districts that had not seen militancy before but have had reserved forests or sanctuaries where poaching had been taking place are found to have been infiltrated by militants later. Consequently, even while there may not be an overwhelming evidence yet, directly indicating links between the two, the inference that the presence of one encourages the other to venture in is prima facie quite obvious.

Figure 21 Districts where poaching and militancy coexist

The map showing the illicit wildlife trade routes from Odisha has been prepared after a field study in course of which information gathered during covert interactions with several players involved in the game has been taken into account. Showing primarily the illicit movement of elephant tusks and tiger parts, the map seen with the previous one showing areas where both poaching and militancy exist puts things in a clearer perspective. Even while trade routes do exist from the western part of the state into neighbouring Chhattisgarh state and from the south to Andhra Pradesh further down south, the length of the affected border with these states in the west and south is relatively less than in the north with Jharkhand and West Bengal. These states have been affected by left-wing militancy as Odisha is.

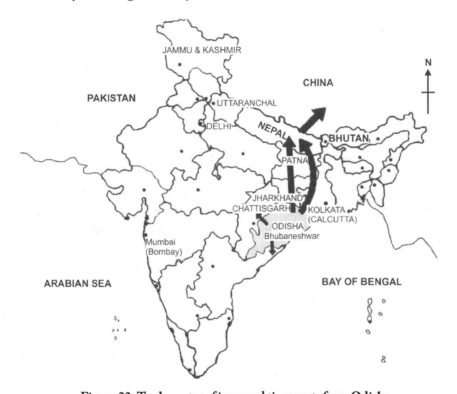

Figure 22 Trade routes of ivory and tiger parts from Odisha

Note: Map not to scale and does not represent authentic international boundaries.

Figures 19 and 20 would suggest that there is not much of an area divide between districts where both poaching and left-wing militancy coexist, as the affected areas were largely contiguous. However, from Figure 21, one could infer that poaching in insurgency affected districts or vice versa is pronouncedly more in the central and northern districts. Trade movement as seen in Figure 22 is largely from these northern districts through states farther north into Nepal.

Even while the focus of this study has been on elephant poaching for ivory, this is not something that can be viewed in isolation. Tiger skin and body parts fetch huge amounts in the black market, and the felines are as much a target of poachers as elephants are for their tusks. There will, therefore, be some attention given to the fate of the tiger in the country and the state as well to understand the issues in tandem.

The illicit trade routes suggest that elephant tusks do make their way into Chhattisgarh on their way to Delhi and to the south for feeding smaller markets. However, poached ivory and tiger parts primarily make their way to Nepal which is an international market feeder. In the movement from the northern districts, the contraband makes its way from Odisha into Jharkhand and West Bengal from where it moves to Nepal for the international market, particularly China.

What is common in the origin and path of the routes adopted for the illicit ivory movement is that the districts from which they originate in Odisha have been severely Naxalite-affected, namely Mayurbhanj, Keonjhar, and Sundergarh in the north and Rayagada and Gajapati in the south. Significantly, in the north, the trade route is through east Singbhum, Simdega, and west Singbhum districts of Jharkhand where Naxalite presence is high. The alternative route to Nepal too is through east and west Midnapore districts of West Bengal where there is significant Naxalite presence, though Nepal itself being a country where Maoists have a significant presence.

On the west, most of the areas of Chhattisgarh adjoining Odisha are severely under the influence of left-wing militancy as also are the districts of Khammam, Visakhapatnam, and Vizianagram in Andhra Pradesh. From their origin in militancy-affected areas, the trade routes pass through similarly affected areas in the neighbouring states and find their way primarily to Nepal where left-wing Maoists had won a huge majority in the country's first election to the National Assembly and despite lesser numbers thereafter, continue to have a major sway at large.

Though no direct evidence by way of specific cases may have come to notice as yet, the circumstantial evidence pointing towards militants being involved in poaching and the illegal wildlife trade or at least of there being links between them in furtherance of this clandestine trade is visible, even if still somewhat translucently. A large bulk of the poaching is happening in forests where left-wing extremists have a major presence, and transportation is also taking place through the left-wing extremist 'red' corridor into left-wing dominated destinations. It would appear improbable that poachers would be operating in forests with such presence completely independent of them. The logical inference on circumstantial evidence is that the trade helps militants keep up their supply of sophisticated arms and ammunition which they do not ever seem to be in short supply of despite large-scale seizures from time to time. Intelligence inputs suggest that a proportion of such arms meant for the Maoist war against the Indian state could well be coming through or from sources in Nepal.

With armed militants having taken refuge in these forests, it has not always been feasible on the part of wildlife officials to adequately patrol such forests let alone take them on. Controlling armed insurgency is the job of the police, not of forest officials. Consequently, the latter do not treat insurgency as their problem even if the militants operate from forests under their control. The police too, whose job it is to confront the militants, understandably does not place poaching very high in its order of priorities. With both militants and poachers operating from the same area, the official disconnect in tackling them separately has had its disadvantages. This likely situation also explains why statistics show that cases of direct poaching have decreased drastically while cases of death from unknown reasons have risen very sharply. To reiterate, reported cases of poaching decreased from 65 in the first period at an average of 13 per year to 42 in the second at an average of 8.4 down to 32 in the third period at an average of 6.4. The total number of poaching cases officially reported in 2015–16 has been just 2. Conversely, during the same periods, death from unknown reasons increased from 15 to 38 to 75 at an average of 3 per year in the first period to 7.6 in the second to 15 in the third. Deaths from unknown reasons for the year 2015–16 alone have been 21.

The IFAW report on 'Criminal Nature' which informs about the increasing role of militants in the illicit international wildlife trade specifically mentions India where elephants and rhinos are targeted for their tusks and horns, and a tiger allegedly killed every day. Even while the tiger statistics may be an exaggeration, the fate that they have been facing in recent times calls for some special scrutiny.

From an analysis of the circumstances prevailing in the state of Odisha, it would appear that the killing of elephants, as also of tigers whose numbers have dropped alarmingly, could possibly be getting a fillip because wildlife officials are understandably hesitant to patrol forests having militants' presence.

Thereby, such killings remain unnoticed for long periods. When the deaths are discovered later from bones and remnants of the elephants, there is no way other than to pass them off as having happened from unknown reasons. In fact, circumstantial evidence suggests that a large proportion of such deaths from unknown reasons are in fact cases of targeted poaching. As the routes through which the ivory is transported also are areas controlled by the militants, the existence of a certain fine-tuned efficiency in the entire illicit poaching and trade operation seems apparent.

Plate 24 Discovered late in dense forest, this could be a case of poaching, but there is no way of knowing; death would therefore be deemed to have been for reasons not known

In addressing the specific issue of poaching and militancy, the study will examine the Simlipal Tiger Reserve in the Mayurbhanj district of Odisha and the circumstances that appear to have facilitated the suspected nexus between them. Even while this case study is based on a specific area, it is intended to be microcosmic and representative of the situation elsewhere too. This tiger reserve which has been a home also to several hundred elephants was hit in a concerted, multi-pronged armed left-wing extremist attack in March–April 2009. Widespread poaching that happened thereafter, when the reserve was virtually abandoned, led to a study team that visited it subsequently terming it as free for all situation.

Notes

1. 'CAG Report Says 56 Elephants Die Annually in Orissa', 17 February 2009, available at http://www.orissadiary.com/CurrentNews.asp?id=10865.

2. *The New Indian Express*, '14 Jumbo Corridors Identified to Reduce Man–Animal Conflict', Bhubaneswar, 28 August 2013; *Odia Sambad*, 'Heat in house on elephants—718 elephant deaths in 14 years', Bhubaneswar, 28 August 2013.

3. Letter no.5129/1WL(H)/4/07, Bhubaneswar, 22 December 2012; no. 4720/1WL RTI 4/2015, Bhubaneswar, 8 June 2015; no. 3647/4WL(G) 4/2015, Bhubaneswar 7 May 2016, Office of Principal CCF (Wildlife) & Chief Wildlife Warden, Bhubaneswar, Odisha.

4. Project Elephant-Ministry of Environment, Forest and Climate Change; envfor.nic. in/division/introduction-4; E-mail with attachment titled 'Total Elephant Mortality in the country' received from *Project Elephant Division*, MoEF & CC, New Delhi by the author at binoybehera@hotmail.com at 1730 hours; E-mail titled 'Death due to various reasons in the country' with attachments received at 1641 hours on 4 March 2016.

5. Letter no.5129/1WL(H)/4/07, Bhubaneswar, 22 December 2012; no. 4720/1WL RTI 4/2015, Bhubaneswar, 8 June 2015; no. 3647/4WL(G) 4/2015, Bhubaneswar, 7 May 2016 of the Office of Principal CCF (Wildlife) & Chief Wildlife Warden, Bhubaneswar Odisha.

6. T. Trautmann, *Elephants & Kings: An Environmental History*, 14; Kautilya, *Arthashastra*, 2.2.13-14, 2.2.15-16, 7.12.22-24.

7. C. Russo, 'Mourning the Loss of a Great Elephant: Torn Ear', *A Voice for Elephants*, *National Geographic*.

8. C. Dell'Amore, 'Beloved African Elephant Killed for Ivory—Monumental Loss', 16 June 2014, *National Geographic*, 14 June 2014.

9. Akshaya K. Patra, Why So Many Elephant Deaths in Odisha?, *Odia Sambad*, Bhubaneswar, 9 October 2010.

10. Hemant Kumar Rout. 'North-East Poachers Sneak into Simlipal, Kuldiha Forests—back in 1994, Lisu Tribals Were Arrested for Mass Killings of Elephants Here', *The New Indian Express*, Bhubaneswar, 14 October 2012.

11. Letter no. 3647/4 WL(G) 4/2015, Bhubaneswar, 7 May 2016, Principal CCF (Wildlife) & Chief Wildlife Warden, Odisha.

12. S. Mohanty, 'Going After Poachers—CB to Probe Jumbo Killings in Simlipal', *The New Indian Express*, Bhubaneswar, 31 July 2012; *Odia Sambad*, 'Crime Branch to Look for Elephant Killers', Bhubaneswar, 31 July 2012.

13. Letter no. 26806/CID-Inv, 28 June 2013, Additional Director general of Police, Criminal Investigation Department, Crime Branch, Cuttack,Odisha.

14. H. Mohanty/*TNN*. 'State Sees a Decline in Tusker Population', *Times of India*, Bhubaneswar, 28 October 2006, Berhampur dateline.

15. *Odia Sambad*, 'Tusker Annihilation in Sorada forests: Nepali Poacher Escapes with 2 Others, One Arrested', Cuttack, 11 August 2013.

16. 'Elephant Deaths Rampant in Orissa', *Press Trust of India* release published in *The Hindu*, Bhubaneswar and other dailies, 31 October 2012.

17. 'Illegal Wildlife Trade Threatens Security', review of the International Fund for Animal Welfare IFAW report; D. Vergano, 'Criminal Nature: Global Security Implications of the Illegal Wildlife Trade', 8, MCT Information Services, *The New Indian Express*, Bhubaneswar, 25 June 2013, available at http://www.ifaw.org/united-states/news/criminal-nature-dangerous-links-between-poaching-and-organised-crime.

18. L. Neme, N. Kalron, and A. Crosta, 'Terrorism and the Ivory Trade', *Los Angeles Times*, 14 October 2013; *National Geographic*, 'Al Shabab and the Human Toll of the Illegal Ivory Trade', 3 October 2013, available at newswatch.nationalgeographic.com; 'Africa's White Gold of Jihad: al-Shabab and Conflict Ivory', available at elephant-league.org/project/africas...ivory/2011-12.

19. Available at https://en.m.wikipedia.org>wiki/Garissa_University_College_attack.

20. Available at news.nationalgeographic.com/news/2014/08/140818-elephsnts-africa-poaching-cites-census/.

21. D. Carrington, 'African Elephants "Killed Faster Than They Are Being Born"', *The Guardian*, 3 March 2016, available at https://the guardian.com/environment/2016/mar/03/African-elephants-killed-faster-than-they-are-being-born, quoting S. John, Secretary General, Convention on Trade in Endangered Species (CITES).

22. A. Vaughan, 'Record Number of African Rhinos Killed in 2005', *The Guardian*, 9 March 2016, available at https://www.theguardian.com/environment/2016/mar/09/record-number-of-african-rhinos-killed-in-2015.

23. *The Times of India*, 'Illegal Wildlife Traders Have Links with Terrorists', Bhubaneswar, 6 July 2013, New Delhi dateline quoting the Union Environment & Forest Minister of India.

24. S. K. Dubey, 'Maoist Movement in India—An Overview', 6 August 2013, available at http://www.idsa.in/system/files/BG_MaoistMovement.pdf.

25. *The New Indian Express*, 'Poachers Have Killed 73 Jumbos Since 05', Bhubaneswar, 29 August 2012 quoting the Minister for Environment and Forest of the Odisha, B. Routray.

26. *Indian Express*, 'Naxalites Terrorists, Reiterates Shinde', New Delhi, 12 June 2013, available at http://www.indianexpress.com/news/naxalites -terrorists-reiterates-shinde/1127853/.

27. Harsha Bardhan Udgata, 'Man-Wild Animal Conflict in Orissa', *Orissa Review*, 67, nos 7 and 8 (2011): 59–63.

28. Ibid.

29. 19 Revenue districts (21 police districts) out of the 30 (32) in Odisha have been declared SRE (security related expenditure) eligible by the Ministry of Home Affairs, Government of India, New Delhi as intimated vide SRE DISTRICTS. docx of Director, Intelligence, Odisha, 4 July 2013.

4

Poaching and Militancy

Originally, an elephant reserve and hunting ground for royalty, Simlipal in the Mayurbhanj district of Odisha was formally declared a tiger reserve in 1956 and was among the first lot of tiger reserves in the country brought under Project Tiger after its formation in 1973.

As per official records, there had been 24 cases of tusker poaching in and around the Simlipal Tiger Reserve from 1990–91 to 1999–2000 at an average of 2.4 per year. In 9 years from 2000–01 to 2008–09, there were 23 cases at an average of 2.6, while in 3 years after 2009, the killings jumped to 18 cases at an average of 6 a year. Incidentally, these figures do not reflect tuskers that might have been killed but did not come to official notice in the aftermath of the Naxalite mayhem of 2009. A fact-finding team that visited Simlipal in June 2010 reported that there were estimated 18 elephant killings in just one month or so prior to its visit.

Simlipal Tiger Reserve came under a series of concerted, well-orchestrated armed left-wing Naxalite attacks in late March 2009, to April, most of which happened over a few days beginning the night of 28–29 March. Seventeen criminal cases were registered in all by the district police. As wildlife officials manning the reserve lost control, the sanctuary where there were already reports of poaching became a veritable haven for poachers targeting its two principal residents, the tusker elephant and the tiger. It was the brazenness of the elephant killings detected in April–May 2010, a year after the attack by militants that prompted the National Tiger Conservation Authority (NTCA) to constitute an independent assessment team in June 2010, to analyse the situation.

The team visited the Simlipal Tiger Reserve from 6 to 11 June 2010, and did an appraisal on the state of poaching in and around the park. Relevant portions of the report will be referred to where pertinent, but the author shall primarily draw on his own experience and association with Simlipal over more than 25 years to come to some understanding and assessment of the situation. This association began in 1988 when the author was posted as Superintendent of Police at Baripada in Mayurbhanj, the district where the Simlipal Tiger Reserve is located.

Simlipal over the Years

Mayurbhanj, the northern Odisha district bordering the states of Jharkhand and West Bengal, was the only 'A' category princely state of pre-Independence Odisha and Simlipal as much a source of forest produce as a game reserve for royalty. *Sal* and other products were transported on the narrow gauge rail line that connected Bangiriposi on the northern fringe of Simlipal to Rupsa in neighbouring Balasore under the then British administration. The only other princely state in Odisha with its own railway was Parlakhemundi, now Gajapati district. For exploitation of its forest resources as also for the hunting convenience of Mayurbhanj's royal family, there was considerable infrastructure within the reserve in the form of rest houses and buildings where these forest offices operated from. Following merger of the princely states with the Indian union after independence, Mayurbhanj became a separate district of Odisha.

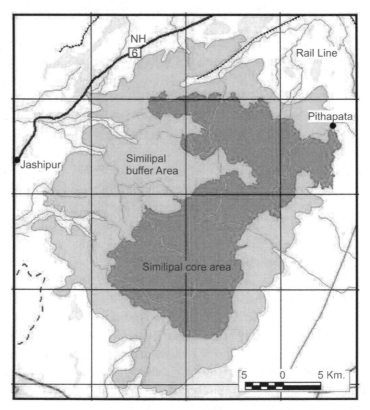

Figure 23 The geographical location of core and buffer areas of Simlipal Tiger Reserve[1]

The Simlipal Tiger Reserve is spread across an area of 2,750 square kilometres of part evergreen and moist deciduous forests and is home to a wide variety of flora and fauna. The name Simlipal is derived from Simuli pahar which literally means the hills with simuli, silk cotton trees. Despite its name, it is *sal* vegetation that predominates, particularly in the dense southern parts of the reserve. The place which has since been declared a biosphere reserve also abounds in ferns and orchids that thrive in the dampness and coolness provided by perennial water sources and the thick vegetation. Orchids include some new species discovered in the reserve, particularly in the Meghasani area of south Simlipal. Its jungles are contiguous with the Kuldiha sanctuary areas of neighbouring Balasore district in the south-east and the forests of Karanjia in Mayurbhanj leading on to those in adjoining Kendujhar district on the west.

Fauna in the Simlipal Tiger Reserve includes predators like the tiger, leopard and small jungle cats, and large herbivores like the elephant and the Indian gaur, the elephants largely are resident of the sanctuary and do not out of it very often. There is also the more plentiful sambar particularly in the southern areas while large herds of chital that were earlier found predominantly in the north have since been seen in the southern side of the reserve. The solitary barking deer and tiny mouse deer are also residents of the reserve. The author who had, sometime or the other, the privilege of coming across most species of major fauna that have been in the biosphere area also remembers having seen packs of dhole, the Indian wild dog that has not been seen in the reserve after 1996. Scavenger vermin like wild boar abound while jackals and the occasionally seen striped hyena are also present though more in the outskirts close to villages. The reserve is also residence to the sloth bear and the king cobra among others, and to over 300 species of birds. And during the monsoon season, leeches are found in millions, living off the blood of animals.

After the first tiger census in 1972 suggested that there were just 1,800 odd of the felines left in India. Simlipal, where they abounded over 1,195 square kilometres of core forest and 1,555 square kilometres of buffer was among the earliest declared tiger reserves following the launch of Project Tiger in 1973. Even though on record the park has had around a hundred tigers, give or take a few, and close to 400 resident elephants, it does not enjoy the full status of a national park because some human habitations continue to exist within. The highest point in the reserve is Meghasani—literally meaning the place where clouds hover—about 3,800 feet above sea level. It is here where the wireless repeater stations of both the state police and of the tiger reserve authority have been functioning.

The report submitted in June 2010, by the two-member assessment team constituted by the NTCA presented a rather dismal picture of Simlipal in the wake of widespread poaching that followed the Naxalite attacks in 2009. The team found animal presence to be scarce, particularly of valued residents like tusker elephants that the reserve had been famous for. The report suggested that animals had dropped considerably in numbers and the picture that came across was that the vibrant wildlife and lively animal behaviour in the park of earlier days had quite changed.

Some of the points which the fact-finding team highlighted were:

- Death of seven elephants in the immediate past was confirmed, all of them most likely caused by poachers.
- Very little animal presence was noted, and not a single tusker was seen during the team's movement and stay for about a week in the reserve.
- Regular incursions into the park by mass tribal hunting groups of 100–200 were happening, at least thrice during the team's stay.
- The honorary wildlife warden confirmed evidence of three more elephant carcasses and information of seven to eight more deaths, bringing the total of elephants possibly shot or poisoned over the previous month or so to 18.[2]

Even while it continues to be an important indicator, the state of wildlife in a particular habitat is not determined entirely by whether the animals there are in adequate numbers. A significant parameter of assessing wildlife well-being in such habitat is the vibrancy of natural animal behaviour. If extraneous interferences cause major changes in the behavioural pattern of wildlife, there are problems to be addressed. In this reserve, apart from the apparent fall in animal numbers, there appeared to have been a significant change from the dynamic spontaneity that existed before to a fearful, self-preserving cautiousness that seemed to have come to prevail thereafter. What the author means by vibrancy of wildlife behaviour may be appreciated from some examples he had come across involving the two major resident species of Simlipal, the two that have been the most affected in recent years.

The tiger, whose presence is an indication that the wildlife environment is healthy, is a fiercely territorial predator. Depending on the prey base available, the extent of territory it marks for itself varies. With chital plentiful in the northern parts of Simlipal, sambar in the south, and wild boar all over, the 2,750 square kilometres that the reserve is spread over would ordinarily support enough prey to sustain a sizeable number of tigers. For the smaller solitary leopard, sambar may be rather large to bring down on its own unless a fawn or subadult and boars sometimes too strong for it, but the tiger being at the top of the food chain and having no natural enemies, these herbivores have been among its chosen prey as the occasional sight of half eaten, putrefying carcasses indicated. However, it is natural for predators to choose options that are relatively less physically demanding and more assured of success. Consequently, the preferred prey differs from reserve to reserve and forest to forest depending on availability, accessibility, and relative chances of success.

An interesting phenomenon that the author had come across in the early days in the reserve was that almost whenever there were small elephant calves around, tigers seemed to be around too, even if not always visible. While sighting of elephants—both solitary tuskers and herds—was normally certain during visits to the reserve, the visibly nervous manner in which particularly small herds tried to shield infant calves in their midst and the often heard growling of the tiger usually suggested its presence nearby. The tiger's growls were almost invariably responded to by the agitated guttural rumbling of the mother elephant, particularly when it was dark and the low visibility necessitated a show of aggressiveness. Even while one could sense the tiger's presence in such situations, there were times when the more confident tigers openly stalked baby elephants in the full visibility of daylight.

For the tiger, targeting small elephant calves involved less effort and greater chance of success than ambushing the fleet-footed chital and sambar or the resilient wild boar. In small herds particularly, calves did not have the kind of protection that larger herds provided. All that the tiger had to do was to locate a small elephant group of three or four with an infant and follow it at a short distance. There was no need for the tiresome efforts at concealment because all that was needed was the moment when the frisky calf would frolic away from the group long enough for the tiger to jump on it. With the tiger's fangs biting into fragile skull and claws tearing into the calf's soft body before the elephants rushed up, all that it needed to do was to keep following at a little distance. It would only be a matter of time before the calf succumbed and its body left behind for the tiger to take over. The calf provided 60 to 70 kilograms of tender meat that would last for a week.

There was this occasion when a particular minister of the state government was visiting the Mayurbhanj district headquarters at Baripada to take the 26th January Republic Day parade salute. Having arrived a day early with the intention of visiting the park, he was accompanied by the author and the park director to Chahala in the northern part of the park. It was around 11 in the morning when three elephants with an infant calf were at the salt lick kicking loose the salt-laden soil and ingesting it.

A slight movement nearby drew attention to something flashy which was seen to be a tiger's coat glistening in the winter morning sun. For the entire duration when the scene was enacted, the tiger had leisurely lain there with no effort to conceal itself and had not taken its eyes off the calf. The herd members themselves did their best to keep it surrounded in their midst. The mother too took pains to see that the calf did not move away from under her belly, all the while emitting rumbling guttural sounds as much to keep the tiger at bay as to warn others of the imminent danger. It is not known what transpired after the herd had left the salt lick but, going by what one had been seeing, the tiger would certainly have followed and waited for its chance.

That was in broad daylight. Coming to night time and the other end of the park late one evening at Upper Barakamuda in southern Simlipal, the repeated growls of a tiger interspersed with answering elephant rumblings had been going on for well over an hour. A little distance down the ravine was where the sounds seemed to be coming from and as the night-time visibility was poor, the author took the help of a torch. The light beam fell on a small elephant family of mother, calf, and subadult walking in a row on an elevated pathway some distance in front.

Avoiding the light on them, the three moved from the pathway into the darkness down the ravine when there suddenly emerged into the beam of the torch an enormous tiger, the tip of its upright tail flailing in its single-minded pursuit of the calf. Even while the author retained his presence of mind enough to keep the torchlight on and slowly retrace his steps backwards so as not to distract the tiger towards himself, the fact that it seemed completely oblivious to his presence only a few yards away was testimony of how completely inconsequential the presence of the author was to the predator.

The two incidents at opposite ends of Simlipal involving the two dominant species of fauna, one during day and one at night, reveal what the author refers to as the natural vibrancy of wildlife behaviour, unaffected by the presence of humans. Going a step further, the incidental presence of humans in a similar incident was looked upon by prey elephants as a source of assistance rather than intrusion. The occurrence was narrated by those who witnessed it.

In the Upper Barakamuda range in the southern part of the park, a tiger that had been following a herd of elephants had found the calf in their midst isolated long enough to attack and mortally wound it. Even as the herd moved slowly through the jungles, the tiger had followed a short distance behind.

Realising that the calf stood no chance if they remained in the jungles, the herd made its way close to the forest ranger's office. The calf had until then walked with some difficulty but having lost a lot of blood, collapsed a little later. The pachyderms hung around for some time after failing to nudge it back to its feet. They then moved away a hundred yards or so and waited, leaving the calf lying there.

Interpreting such behaviour as a signal for their intervention, the range officials got out with their disinfectants, their wads of cotton and used them on the bleeding calf before returning to their office. The herd returned and the mother again tried to nudge the slumped body back on to its feet. Passing their trunks over the calf as if to feel the lifeless body for the last time, the herd did not walk away from it as one would have thought they would. They hung around making low rumbling sounds as if conversing and were communicating with other elephants in the jungle. It must have been through low frequency sounds, not very audible to human ears, and the occasional trumpeting.

It was not long before a few other herds made their way to the spot from different directions. After congregating near the dead calf and hovering together through trunk contact and rumblings as if in commiseration, 20 to 30 elephants moved away some distance and waited. Some of the staff who witnessed the incident and who this author spoke to said that they felt as though the elephants wanted them not to let the tiger claim their calf. The pachyderms kept their distance as the officials dug a pit close to where the calf lay. Even while acknowledging that they were interfering in nature's predator–prey relationship and unfairly depriving the predator of food earned in the due process of nature for its own survival, they said they felt vindicated when the elephants finally left only after the calf had been buried.

The purpose of narrating these incidents is to illustrate how in a healthy wildlife environment, animal behaviour can be so vibrantly spontaneous and unaffected by human presence. In a changed environment, the reason why the NTCA-assessment team could not see any tuskers during its five to six days' stay in the park could well have been because there were not many of them left to be seen. It is also possible that those that might still be there chose not to expose themselves because in the backdrop of heightened poaching, their survival instinct warned that exposure would make them vulnerable.

Elephants are intelligent animals with long memories and, given the strong social bonding that they have, they apparently have ways of communicating through means not entirely known to us. The experiences of poaching, particularly in recent times when some elephants might have escaped but several killed would have made them wary. The author too, during visits to Simlipal over some years, had not come across too many of them and even when he did it was only in the very late hours of the night. Earlier, elephants were seen at all times and all places. Their changed behaviour had apparently been conditioned by man's intrusions because of the threat they saw from him. While sighting of reasonably large herds of individuals was fairly common, well-endowed, large tuskers too were also plentiful. Even while the size of the tuskers and their tusks were generally bigger, their confident behaviour too was markedly different from the fearful demeanour of self-preserving concealment that came to be seen thereafter.

There was this gregarious but temperamental tusker at Joranda, a handsome fellow with well-formed tusks who did not like to be dictated around. As at all offices and rest houses in the Simlipal reserve, there was an 8–10-foot-deep and about 7–8-foot-wide trench around the Joranda complex as well. To allow vehicles to enter, there were four *sal* logs embedded into the soil at opposite ends of the trench at right angles to it where the road entered the complex. Wooden planks were placed across these logs to allow vehicles to enter, and the planks were removed at night time to keep elephants out. Even while smart and determined elephants had no problems balancing and walking the length of the *sal* logs into the premises, this particular tusker made it a point not only to balance through but also to fastidiously place the wooden planks that had been removed back on the logs before leaving. And if the officials persisted with removing the planks at night time despite his obvious disapproval, he would do a little damage to the office building or the rest house to express his annoyance. 'Don't even think about trying to keep me out, this is my area,' he seemed to be saying. Interestingly, when the planks were later left on and not removed during night time, he seldom bothered to enter.

Then there was this massive tusker with thick and shapely tusks lording around the Chahala area. In a daytime situation near the range office, while observing a subadult male at the salt lick from the safety of the trench in between, the giant made his entry. Even as the elaborate process in which the subadult crouched on its forelegs, used its growing tusks to dig up the soil, straightened up, kicked the earth loose with its forelegs, and then used its trunk to put portions of the mineral laden soil into its mouth was being enacted, someone around announced a little loudly that another elephant had emerged from the forest.

The enormous tusker stopped, slowly looked around towards the source of the sound, and changing direction, walked with cool confidence to the edge of the trench on the other side of which we were. He stood there for a long time majestically staring down at us from 10 to 12 feet away. If ever the author thought he could understand elephant communication, it was then. The tusker bull was clearly saying this was his turf and if intruders wanted to visit, they should not be disturbing him. Having stood long enough for us to get the message, he turned around and walked regally to the saltlick.

There have been several fascinating interactions with single tuskers and herds revealing various aspects of elephant behaviour in Simlipal that the author feels tempted to narrate. However, the purpose here is not to romanticise wildlife through dramatic storytelling but to illustrate through examples from different areas, the all-pervading ambience of this vibrant and uninhibited wildlife behaviour. Elephants can live up to 70 years but the only remnants that would be left today of the Joranda boss, the Chahala lord and many like them would probably be in ivory carvings showcased in the drawing rooms of some affluent connoisseurs somewhere.

Poaching

Wildlife researchers have opined that villagers living around Simlipal who had served in the defence and paramilitary forces in the north-east had developed contacts with locals there who were adept at hunting elephants with poisoned arrows. Seeing the ivory poaching potential in Simlipal, they liaised with them for the spate of killings that started in the reserve in 1994. Even though killing initially began outside, it was soon established in a big way inside the reserve with specific places identified by ivory poachers within the park for concealment. Trade-offs were agreed upon outside and delivery was made directly to purchasers from the places where the tusks were hidden.[3] Intelligence inputs and investigation into some of the cases have confirmed the active involvement of Lishu tribals from Arunachal Pradesh and Mizoram in killings that have not only continued unabated but have also spread to other parts of the state as well.[4]

An important object of this study has been to try and address some fundamental issues that led to the situation where poachers came to have such sway over the park that not only tuskers disappeared but also tigers did in large numbers. The question that begs to be asked is what the circumstances were that facilitated this sudden, sustained burst of poaching when the villagers who are said to have been engineering poaching earlier had been there around 1994 too. More importantly, what was it that led to such a state of affairs that armed militants came to have complete control of the Simlipal Tiger Reserve in the latter part of the 2000 decade?

In objectively trying to answer some of these questions and arriving at possible conclusions, the author does not wish to specify individuals on record. However, in order to establish that important decisions with far reaching consequences need to be taken in a systemic manner rather than on personalised orientation, it is necessary to highlight certain individual issues which, even while meaning well and purportedly taken for the well-being of wildlife, could be counterproductive in the long run.

Akhand Shikar—literally meaning uninterrupted hunting—has been a mass hunting tribal ritual in Simlipal from the times the Raja of Mayurbhanj ruled the princely state. It was programmed to coincide with the festival of Chaitra Parva, generally around March–April each year when the signature martial dance of Mayurbhanj, Chhau, was among the various customs also celebrated. Chaitra Parva was also the week of festivity when the Mayurbhanj king permitted *akhand shikar* in Simlipal for the indulgence of his subjects. As tribals constituting the vast majority of the king's subjects and most of those who went hunting were from among them, *akhand shikar* came to be recognised over the years as a tribal custom.

Even while the glorification of Chhau during Chaitra Parva has continued to this day, *akhand shikar* became illegal once hunting was banned much like it has been with Maasai tribals in Kenya whose custom of spearing a lion and bagging its tail as a test of youth prowess was brought to an end. However, *akhand shikar* remained in the simplicity of the tribal psyche more the continuance of an old custom than an illegal act of mass defiance, and they sometimes insisted on being allowed to kill a few animals as token and distribute small quantities of the salted meat in observance of the tradition.

Park Management and District Police

During the years when the author was a Superintendent of Police in the district over the last few years of the 1980s decade into the early 1990s, elaborate police arrangements used to be made to prevent akhand shikar hunting and apprehend hunters not just in the vulnerable areas of the park but in entry points as well. Simultaneously with the Simlipal wildlife authorities' own anti-poaching arrangements with the resources available with them, due coordination was maintained between the wildlife and security enforcement agencies with officers in charge of police stations overseeing police arrangements in areas of the park under their jurisdiction. The deployment remained in place usually for the week or until such time as the mass hunting came to an end.

In a stark reflection of the changed situation, *akhand shikar* that used to happen on a fixed time schedule once a year was reported by the NTCA-constituted team in 2010 to have become a routine. Three forays into the park by armed tribals numbering 100–200 were reported in the 5–6 days that the team spent in the reserve in the month of June. In other words, frequent instances of mass poaching by large groups of people had begun in the park, and despite being called *akhand shikar*, this was different from the astrologically timed, traditional tribal custom of yore.

This is not to suggest that there was no poaching in the area then. But the intelligence collection network was efficient and strong coercive action was taken immediately on something adverse coming to notice. One of the earliest known mass buyers of elephant tusks was an otherwise reputed businessman who was arrested after his clandestine activities came to light in 1990–91. His apprehension and recovery of ivory from his possession were as much a result of intelligence inputs from the police as proactive action by wildlife officials.

However, poaching was limited and Simlipal thrived, a major deterrent for detractors being the publicly visible coordination that existed between the police security mechanism and the wildlife enforcement apparatus. The frequent presence of the author himself as head of the district police force and of field police officers in their respective jurisdiction within the park ensured timely collection of field intelligence as much on poaching as on possible crime and potential law and order issues.

A major problem then confronting the district police and administration was the violent agitation instigated from across the Bihar state border demanding that Mayurbhanj district secede from Odisha and become a part of a new state of Jharkhand. There were occasions when agitators with weapons took refuge in the inaccessible areas of the sanctuary, but their plans were usually thwarted because field police officials often visited the reserve and were able to obtain intelligence for timely preventive action. Simultaneously, if information on any congregation of poachers came to notice, it was passed on to the park authorities for necessary remedial action. Intelligence gathering in the park was as much from the many villages within the reserve and its outskirts as from evidence of gathering seen inside the reserve or assessed from the remnants of cooking arrangements and meals eaten.

Deserved credit to the park management of the day is due for the rapport they maintained with the police which helped both fulfil their respective responsibilities and duties efficiently and effectively in the interest of both state and society. More importantly, the message of this understanding was not lost on the criminal sections of the public inclined to break the law or resort to poaching. It may be highlighted here that the check on poaching was possible despite the rather inconvenient administrative structure of the forest establishment where the Park Director's jurisdiction was confined to the core area of the sanctuary while the buffer region was under the control of another authority called the Simlipal Forest Development Corporation. The territorial Divisional Forest Officers outside the reserve in turn reported to yet a separate Conservator of Forests located several hundred kilometres away at Angul.

Even while the integrated intelligence and enforcement network of the police and wildlife authorities continued to work well for some time after the then incumbents of the respective departments had moved on from the district, signs of strain started emerging when subsequent park managements showed a disinclination to allow government officials other than their own into the park. The tiger reserve authorities would surely have had their reasons for doing so, maybe because of inadequate level of understanding with and cooperation from the police and the district administration. However, instead of nipping it in the bud, the shift in focus from what had helped systemic interests before to what was beginning to happen did not augur well for either of the two enforcement systems that had been working in tandem to their mutual benefit.

Though the disjointed administrative structure in and around the Simlipal wildlife establishment had become better streamlined, a rigid interpretation of wildlife law made matters difficult when during the incumbency of a particular park head, even the District Magistrate and Superintendent of Police of the district were prevented from entering the reserve unless they obtained entry permits from an Assistant Conservator of the park. Though Section 38V(4) i of the Wildlife (Protection) Act, 1972, gave park authorities the discretion to keep core areas inviolate so as to prevent disturbance, Section 27 of the same Act that regulated entry into sanctuaries stipulated in subsection (a) that public servants on duty do not need permits to enter any sanctuary. This would perhaps suggest that neither the District Magistrate nor the Superintendent of Police was aware of the provisions and interpretation of law relating to their authority in their own area of jurisdiction.

Prevention of crime in its jurisdiction is the statutory duty and responsibility of the police who as per law are on duty 24 hours a day. Even though Simlipal was not outside the jurisdiction of the district police or purview of criminal laws of the land, when locked gates obstructed even field police officials from entering and park gate officials informed that they did not have the authority to open the gates, a process of official disconnect started at the highest level down the ranks which had ruinous consequences, both for wildlife management and crime control.

When a meeting between the park authority with the district administration and district police, to try and salvage the deteriorating situation, did not yield any reasonable understanding, the countdown for disaster had begun. Personal issues had apparently gotten the better of the larger interests which could well have been addressed and sorted out at a higher level of the administration. But they were not, leaving a heavy price for the state to pay.

The developments, as had been happening had predictably, reached public eyes and ears, and the consequences were along expected lines. With resources available with the park authorities being rather limited for effective vigil on such a large reserve and cooperation of the district and police administration no longer forthcoming, poachers moved in a big way. Local gangs backed by Lishu tribals from the north-east had already started taking a heavy toll on tuskers by the mid-1990s, but by the latter part of the decade, syndicates had entrenched themselves firmly for organised poaching not just of tusker elephants but as has subsequently come to be known from the dwindling tiger population, of tigers as well.[5]

The Tiger

While referring to tusker killings, it is necessary also to make a reference to the tiger to understand the sway that poachers had begun to have in Simlipal. It has already been mentioned in this chapter that the author experienced the presence of tigers during his early years of acquaintance with the park. Actual sighting may not have been very frequent because of reduced visibility through the undulating topographic terrain and thick foliage, but the numerous signs of its presence were unmistakeable. Fresh scat on the roads, remains of kills, frequent alarm calls of chital herds gazing in a particular direction, hoof stomping and raised tails of sambar looking in the direction of the predator, frequent growls particularly at night time, and the typical nervousness of small elephant herds with calf indicated that the cats were around and thriving. Tiger census figures of the earlier century had never suggested that the feline could be in trouble, and even the first census officially done by the state wildlife authorities this century, indicated the presence of 192 tigers in the state. Over a hundred of them were found to be in Simlipal, among the oldest declared tiger reserves in the country.

However, warning signs had already become visible. The decreasing frequency of indications of tiger presence suggested that its population was falling. In course of a tiger census operation that the author attended over a decade and half after having left Mayurbhanj district, no pug marks on the pug impression pads were detected in the northern areas of Simlipal including the once favoured Chahala range. Even while pug mark census is not a very accurate method of ascertaining tiger population, the complete absence of pug marks confirmed that there were no longer any tigers in the northern areas of the reserve.

With reduced predation, it is possible that the expanding prey base of chital in the north was spilling over to southern Simlipal where it had not been seen before. However, here too recent reports have been worrying. Scientists from the Wildlife Institute of India warned some time back that the prey base in Simlipal had fallen so low that having a sustainable habitat for tigers was getting to be difficult.[6]

Significantly, it was not much later that the NTCA that had evolved in 2005 from the original Project Tiger broke the alarming news in 2006 that the number of tigers in the country was the lowest ever in recorded history, just 1,411.[7] Not coincidentally, this coincided with the peak period of militancy and insurgency around the country. Even though the last pug mark census conducted by the state wildlife authorities in 2004 put the number of tigers in Simlipal at 101 out of the total of 192 in the state,[8] the camera-trap method adopted by the Wildlife Institute of India some years later suggested that the actual presence of tigers was much lower.

In the estimate of the NTCA, there apparently were just 45 tigers in the entire state of Odisha in 2006. In the census done by NTCA in 2010, there were estimated 1,706 tigers in the country, while in 2014, the number reportedly rose to 2,226. For the corresponding years, the tiger population in Odisha reportedly fell from 45 in 2006 to 32 in 2010 to just 28 in 2014.[9]

In the war of words that erupted between the state wildlife wing and the NTCA about the relative efficacy of the two census methods and the authenticity of their respective figures, some crucial issues were lost sight of. As per the latest census figures officially released by the Government of India, the NTCA puts the tiger count in Odisha at 28, down from 32 in 2010 of which, in the immediate aftermath of the Naxalite attack, there were estimated 23 tigers in Simlipal and 8 in the Satkosia Tiger Reserve. The state wildlife wing that had in 2014 claimed that there were actually 45 tigers in the state has since claimed there are 40.[10]

This brings up an essential question. Presuming that the figure put up by the state wildlife wing is authentic as claimed, it still does not negate the fact that in course of just one decade the tiger population in the state has fallen from 192 in 2004 to 40 presently. This marks a drop of nearly 80 per cent or close to 4 out of every 5 in course of just 10 years. If one goes by the figure of 28 projected by the NTCA, over 85 per cent of tigers have disappeared in course of the last decade. Even presuming that all the 40 tigers claimed by the wildlife wing to be presently in the state are all in Simlipal and none anywhere else in the state, the drop is seen to have been over 60 per cent in one decade, not an enviable situation for one of the oldest and highly protected reserves of Project Tiger.

What then has made these animals so vulnerable even in such a protected environment? Unlike poached elephants whose bodies are left behind to be found even if belatedly, tiger carcasses are usually not discovered at all. From skin to claws, skull to bones, canines to even whiskers, every part of the tiger's body fetches enormous prices in the international black market and there is very little that is left of the poached tiger to be discovered.

It is no secret that the most favoured destination for tiger parts is China where ground tiger bones are used as ingredients in a variety of traditional medicine from pain balm to aphrodisiac. China's efforts to get the international community's approval for extensive tiger farming have been steadfastly resisted notwithstanding the fact that tiger farms already exist in considerable numbers in that country. Some south-east Asian states like Laos, Vietnam, and Thailand too have their captive tiger farms to meet domestic requirements, but it is in China that the demand far exceeds supply.[11] Even while wildlife authorities in Thailand have come down heavily on tiger temples indulging in tiger parts trade, there is considerable domestic patronage of the trade in China. Such additional requirements as are needed for this huge cottage industry are consequently met from the international black market. India being a neighbouring country and having a sizeable tiger population has naturally been more vulnerable and affected.

With 70 per cent of wild tigers on earth believed to be in India,[12] the felines will continue to be targeted and their parts sent along with poached ivory to China through Nepal. Northern districts like Mayurbhanj having had a sizeable tusker and tiger population to poach, the extensive common border with states like Jharkhand and West Bengal makes it an important starting point of the trade route to Nepal as also a congregation point within the state for poachers to send their wares from.

Significantly, despite the NTCA having announced an increase in the tiger count across the country on the basis of its census done in 2014, their numbers in neighbouring Jharkhand on the north and Andhra Pradesh in the south have reportedly fallen as they have in Odisha. Not insignificantly, tiger mortality around the country has risen sharply in the years since then as also has been the proportion of poaching deaths.

Organised Poaching and Militancy

In the study of possible links between poaching and armed militancy in the Indian context, it would be desirable to know how and why this situation came about in countries of Central Africa in the first place.

It is known since long that drug trafficking has been the major source of terrorist funding around the world and heroin from Afghanistan accounts for 80 per cent of the trade worldwide and raw opium almost 70 per cent. Over the past decade and more, increasingly greater drug supplies from Afghanistan shipped from Pakistan were reaching Africa, particularly countries of the sub-Saharan central part of the continent. This was a major source of terrorist funding in and around these countries, this area called the Sahel belt having become the new terrorism hotspot. Weak drug laws enforcement had helped an extensive trafficking network to develop.[13]

Following proactive involvement of the United Nations Office on Drugs and Crime (UNODC) in drugs law enactment and enforcement in these countries, trafficking was badly hit and many of the lead traffickers jailed for long terms.[14] Even while drug money still continued to flow into the terrorists' coffers, it was much less than before. This was when terrorists turned to large-scale poaching as an alternative source of financing which, even if considerably less than drug money, became a complementary resource nonetheless.

Drug money plays an important role in funding terrorism in India too, particularly in areas like Kashmir in the north and the states of the north-east. In many states affected by left-wing extremism as in Odisha, thousands of acres of cannabis grown on inaccessible government owned or forested areas of some sparsely populated districts have been periodically and systematically destroyed. This has been particularly visible in hilly and rain-beaten tribal districts like Kandhamal that have been witness to both poaching and left-wing militancy. The specific ownership of such large-scale plantations has not been established with any great degree of certitude and is believed to have been of left-wing extremists. It is quite possible that consistent steps to destroy such widespread plantations may have led them to poaching as an alternative, though not known as yet to be in a manner organised enough to compare with Central Africa, nor with the kind of returns there.

The first decade of the current century spilling into the earlier part of the second has been the period when armed insurgency in various states of the country, left-wing militancy in Odisha included, were at their peaks. This has also been the time when the tiger population, mostly in those very states that have seen militancy, has fallen and elephant tuskers and rhinos targeted. Indeed, the announcement by the NTCA in 2006 of the tiger population being the lowest ever came right around the time when armed militancy had reached its most violent summit phase.

In the light of reports that organised poaching is linked to militancy in Asia as well,[15] the disappearance of tuskers and tigers as well as of rhinos in the north-east takes on a new dimension. The investigation report had mentioned that in India, tuskers are being slaughtered for ivory and tigers being killed at will, and the trade used for procurement of arms for perpetrating militant warfare.

The well-protected Simlipal Tiger Reserve in Odisha provides a microcosmic examination of what could possibly go wrong in the security and wildlife management apparatus. Circumstantial evidence cited earlier in the study has suggested that there are links between poachers and militants with poaching being high in many Naxalite-affected districts. If the falling population of tuskers and tigers and the course of events that followed are any indication, the nexus of poaching syndicates and militancy seems to have been active in Simlipal.

There already was evidence of sporadic Maoist militancy presence earlier in the forests outside Simlipal, but in the latter part of the 2000 decade, it had become more pronounced. Between mid-May 2006 and mid-March 2009, there were 24 incidents of armed left-wing extremist violence in the district reported with the police. Not insignificantly, the last such case registered was on 15 March 2009, in the forests outside the park's outskirts under Mahuldiha Police Station, two weeks before the first phase of attacks inside the tiger reserve.[16]

As subsequent developments have proven, simultaneously with organised poaching that was happening and had taken strong roots in the park, left-wing militants had access into Simlipal and had familiarised themselves with the terrain and layout, particularly in the core area. Since the armed Naxalite attacks in the park took everybody by surprise and the familiarity of the militants with the reserve could not have happened overnight, circumstantial evidence points conclusively to their presence inside the park over a period of time during which there was no intelligence input on their activities. As mentioned earlier, the access that the district police had into Simlipal in earlier years had not quite remained that way. Even while the tiger reserve authorities had declared the core area of the park inviolate and purportedly kept it out of bounds in the interest of animal protection, they were clearly unable to detect, let alone prevent the massive infiltration by insurgents into the park.

In the assessment of the author as a professional policeman, there was a reason for the police not to have been adequately aware of what was brewing inside the sanctuary simply because they were removed from it physically. The park management and wildlife personnel working in the reserve should have known, but they do not seem to have had any inkling of what was developing right before them. In any case, checking militancy does not come within the sphere of duties of the wildlife establishment even though the situation had a direct bearing on the well-being of wildlife therein.

The violent consequence of official apathy and resultant Naxalite mayhem suddenly exploded inside the park without the police having much chance to know about it, let alone prevent it. Thrown into a major armed battle, lack of police acquaintance with the sanctuary terrain from years of being denied access made the task of flushing out the militants much more difficult. On the other hand, the obvious familiarity that the militants themselves had with the park terrain only compounded the problem.

Not surprisingly, when the four-member committee set up by the NTCA to assess the field situation after the Naxalite attacks visited Simlipal, the Superintendent of Police of the district expressed strong reservations he had with the wildlife authorities. The committee report mentions that on 3 August 2009, when they met, the Superintendent of Police informed that lack of coordination between the wildlife department on the one hand and the civil administration and law enforcement authorities on the other had made the task of taking on the militants in the forests much more difficult.[17] What the Police Superintendent did not quite explain is what prevented him, despite how the park management might have acted, from taking stronger steps to prevent anticipated crime in his own jurisdiction as he deemed necessary with the statutory powers already vested in him. On the other hand too, this point mentioned by the NTCA team does not appear to have been given much importance either.

Figure 24 shows the places in the tiger reserve that came under heavy Naxalite attack, all within the space of a few days. What comes across noticeably is the widespread nature of the attacks, covering the entire geographical area of the sanctuary and leaving no important part of the core area untouched. The targets chosen and the order in which they were attacked confirms not just the familiarity the armed militants had with the Simlipal Tiger Reserve but the elaborate strategic planning that went into it.

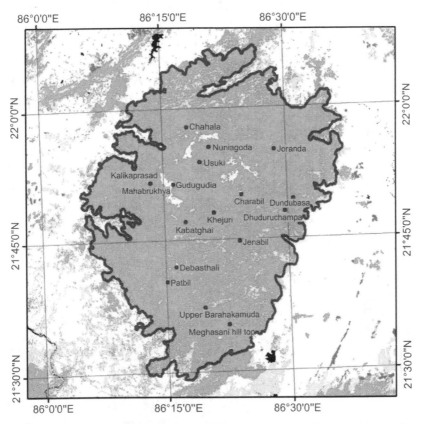

Figure 24 Places in Simlipal Tiger Reserve that came under Naxalite attack in March–April 2009

For maximum effect on morale and material damage, the first hits during the night of 28 March 2009 were on Chahala, the most identified tourist destination of Simlipal, and on Meghasani at the other end, the hill top where the wireless repeater communication stations of both the forest department and the police are located. Dhudruchampa, which was also attacked the same night, seems to have been chosen because of its geographical proximity and assessment by the armed insurgents that breaking themselves up into groups too far apart would increase their vulnerability. The next night, on the 29th, the key Upper Barakamuda range office, Patbil, and the large meadow post at Debasthali were attacked, while on the night of the 31st, the targets were Nuniagoda, Joranda, Gudgudia, Jenabil, and Dundubasa. Usuki was attacked during the night of 1 April; the entry gate to the park at Kalikaprasad and Khejuri on the 3rd, and the incidents of firing and attack at Mahabrukhya, Charabil, and Kabatghai happened thereafter. The planning and coverage were systematic and the execution precise.

Even while the entire park management abandoned the reserve and fled, it was left to the police that had earlier been restrained from entering to step in and take charge. Seventeen criminal cases of Naxalite attacks inside Simlipal were registered, and over four years thereafter when the park was flushed free of insurgents, it ensured security of the reserve by extensive presence of armed police contingents inside. Seventy-eight accused militants were arrested during this period to stand trial in the court of law.[18]

During interrogation of some of the leaders after their arrest, some interesting facts emerged. It appeared reasonably obvious that the top-level planners of the simultaneous, concerted attacks could not have been at the level of those arrested because such perfectly orchestrated, meticulously synchronised attacks had far too much of sophisticated planning for those arrested to be credited with. Following questioning on links with poaching, the answers of some were revealing.

One from among the leaders arrested said that on two occasions he had got directions to receive the cut-elephant tusks weighing about 25 to 30 kilograms immediately outside the reserve at Bangiriposi, and hand them over near the Saranda forest close to Chaibasa in east Singhbhum district of Jharkhand. He claimed ignorance about where or when the ivory had been procured, but said that he was given to understand that like him, the person he handed them over to would help transport them to Nepal. Apparently there were several conduits on the way through Jharkhand and Bihar into Nepal, as the break-up of carriers in transit being a part of the strategy to keep the trade route safe. If none knew about the route the others had taken, all he would be able to expose in the event of being caught would be confined to the route that only he himself had used. Apparently they availed whichever kind of transport was convenient but often used jeeps with cavities created under the seats to carry the contraband.

Another leader separately entrusted with a similar task was more forthright in acknowledging that he and two elephant poachers, whose names he said he did not know but could identify by face, were involved in the killing of tusker elephants in Simlipal. This was in the aftermath of the militant attacks. When the reserve had been virtually abandoned, poaching, extricating tusks, and transporting them out of the park, did not pose much difficulty. A third informed that he had been directed to a particular seasoned poacher from whom he was to collect and deliver certain goods at a pre-decided place some distance into the Purulia district of West Bengal. He was entrusted with two gunny bags which were light for their size but quite voluminous, the rattling suggesting they contained dry tiger bones. These had come from tigers in Simlipal around the same time, and the quantity suggested there were more than just a few that had been killed.

This raises a serious issue. Even while the some of the militants interrogated suggested links with poachers, it was apparent that this possibly was only the proverbial tip of the iceberg, the volume and dimensions of the portion beneath the surface as yet uncertain. There were reasons to believe that extremist masterminds at a higher level called the shots and organised matters, and issued instructions for local level leaders to carry out. If they were doing so at Simlipal, they could well have been organising similar acts elsewhere too.

The NTCA constituted fact-finding team that visited the park some time later in 2010 and had opined that poaching of animals had become so rampant in the park that venison was being sold in the neighbouring villages at rates well below the price of mutton.[19] Even while militants continued to take shelter in the reserve, poachers operated with impunity which is all the more reason to suspect that there has been collusion between them. Mayurbhanj was among the foremost districts in elephant tusker and tiger poaching, and being densely forested, Naxalite infested, and bordering similar forested districts of Jharkhand and West Bengal in the north, the militants–poachers combine clearly extended beyond the state.

With all the three states north of Odisha, namely, Jharkhand, West Bengal, and Bihar, being part of the Maoist-red corridor, proposed to be set up by left-wing extremists, and Nepal being a veritable haven for Maoists, inference on the possibility of deeper links is inescapable. Not insignificantly, the tiger population in Odisha had fallen substantially by 2010 and even further by 2014. As per the last NTCA count, tiger numbers in adjacent Jharkhand have fallen to a mere three while West Bengal still has 79 because most of them are in the inaccessible mangrove forests of the Sunderbans. In Andhra Pradesh too, the state immediately south of Odisha, the tiger population has dropped.[20]

It is relevant to mention the situation in some of the other wildlife reserves and sanctuaries of the state in this regard. Tusker poaching in various districts has already been deliberated on. Coming to the other major denizen of Odisha's forests, the tiger has traditionally been found in most jungles of Odisha, the Satkosia and the Sunabeda Reserves both having been declared tiger reserves after Simlipal. Various forests of Odisha have had tigers until not very long back but they are now confined to just over a score in Simlipal and possibly a few in the two other tiger reserves, Sunabeda and Satkosia. They have all but disappeared from the rest of the state including the Karlapat sanctuary where they once were in large numbers.

Militants have infiltrated just about every wildlife abode and sanctuary in Odisha, except a few like Bhitar Kanika National Park where estuarine crocodiles abound. Unlike the Sunderbans in West Bengal and Bangladesh where tigers thrive in similar wet mangrove forests, there are no tigers here or elephants. The sanctuaries and reserves have since become shelter places for militants or thoroughfares for their movement. Presence of armed militants in established wildlife areas has understandably affected the fate of those animal species that fetch high prices in the black market, not to mention the security threat they generate. Circumstances have apparently been allowing militants and poachers to become partners in crime in a symbiotic relationship of mutual give and take. Prima facie, the presence of militants directly offers protection to poachers or does so indirectly by instilling fear among officials from venturing into such areas, thereby allowing poachers to feel safe.

Be it the wildlife trade route out of the state or procurement of weapons, Nepal as a destination is part of the geography involved. The concept of the Red Corridor dividing India into East and West, envisages Nepal in the north through eight states down south that include Bihar, Jharkhand, West Bengal, and Odisha. In this kind of situation, where just about every wildlife sanctuary spread across the state is affected, the presumption that wildlife trade can no longer be treated as just an environment matter takes on added significance. This is because the issue of wildlife trade has gotten intertwined with the concern of state security.

Anti-poaching Measures and Insurgency

In some articles published by a few credible English and vernacular dailies some time back, the Simlipal Tiger Reserve authority was quoted having said that following restoration of normalcy in the park, elaborate anti-poaching arrangements as are required have been put in place by the park management and no vehicles other than those of the forest department will be allowed inside the sanctuary. Reports also indicated that orders have been passed by the wildlife authority not to allow anyone other than forest officials into the core area of the reserve.[21]

A request letter to the Simlipal Tiger Reserve authority for confirmation of such orders said to have been passed elicited no response but the Chief Wildlife Warden, Odisha to whom the Simlipal authority reports informed that as per Section 38V(4)(i) of the Wildlife (Protection) Act, 1972, core or critical tiger habitat areas of national parks and sanctuaries can be kept inviolate for the purposes of tiger conservation.[22] In accordance with such orders, other officials and their vehicles too would no longer be allowed inside the Simlipal Tiger Reserve.

As had happened before with disastrous consequences for both wildlife and state security, such orders of the Chief Wildlife Warden recall the aphorism that those who forget history are condemned to repeat it. If such norm is replicated elsewhere, it would have grave security implications for the entire country. The accountability as well as the legality of such orders need to be examined as much in the context of Simlipal and Odisha as also with reference to the overall left-wing extremism scenario in India. And the vital question that inexorably asks itself is whether, after a series of Naxalite attacks devastated Simlipal and painstaking efforts brought back normalcy, history will be allowed to be repeated through the same mistakes that allowed it to happen in the first place.

Prevention of poaching is the responsibility of wildlife authorities, and the park management is within its rights to make arrangements for its prevention in the sanctuary. However, the moot question is whether executive orders passed by the park management can extend to a situation where the police is restrained, partly or wholly, directly or indirectly, from discharging its own statutory responsibilities in its own jurisdiction because it happens to be a wildlife sanctuary.

Criminal laws of the country do not apply any less in sanctuaries than they do elsewhere and it is the statutory duty and responsibility of the police to prevent the commission of cognisable crimes in its jurisdiction, wildlife sanctuaries not excepted. If the police is put at par with others and restrained from entering a sanctuary on the grounds that wildlife could be disturbed, collection of intelligence relating to insurgency, a key requirement in all matters of security at both local and national levels, will be severely impaired. This will prevent timely action which will encourage crime, more importantly the heinous crime of waging war against the state that is the avowed objective of left-wing extremists. In the backdrop of the Naxalite attacks that crippled the park for years not very long back, it does not need reiteration that such a situation could happen again.

Studies, including the one done by the Institute of Defence Studies and Analyses, have established that Naxalite insurgents are spread over a major portion of the country's forests.[23] History has also shown that wildlife reserves and sanctuaries are potential strongholds for armed militants because there is relatively less extraneous intrusion for them. After Simlipal in Mayurbhanj, the Sunabeda Tiger Reserve in Nuapada in the state of Odisha has gotten out of bounds because of extensive militants presence there. The third tiger reserve, Satkosia in Angul, which had been free until recently has also become a haven for militants. The Usha Kothi wildlife sanctuary in Deogarh district, the Debrigarh sanctuary in Bargarh district, and the Karlapat sanctuary in Kalahandi district are all shelter places or thoroughfares for armed militants. Surely, the fact that militants have infiltrated just about all wildlife sanctuaries cannot be entirely coincidental. On the contrary, it is more likely than not that they have been directly or indirectly facilitated.

In the aftermath of the Naxalite attacks and restoration of normalcy in Simlipal after years of effort, if the police is again restrained from entering the park, the vulnerability to a repeat situation would be obvious. Social movements do not just disappear, and they reappear in other guises. The important question that comes up is who will be responsible if militants start building their presence in the park again and the police is unable to detect it, let alone prevent it as has happened before.

If the park authorities decide that the core area has to be kept inviolate, it is also their responsibility to ensure that it remains that way, be the potential infiltrators common men, poachers, or militants. While they surely are equipped to take on the first two kinds, it is unlikely that they can do the third, more so because taking them on does not come within the sphere of their duties. Ironically, when the wildlife and security managements should be pooling resources to fight this illegal joint menace, it is their adversaries, the poachers and militants' who seem to have gotten together and one-upped them.

Legal Issues

There are some important legal questions that come up in this regard. Section 27 of the Wildlife (Protection) Act, 1972, provides under 'Restriction on entry in sanctuary' that '(1) No person other than, (a) a public servant on duty (and four other categories mentioned under subsections b, c, d, and e) shall enter or reside in the sanctuary, except under and in accordance with the conditions of a permit granted under section 28'. It therefore follows that a field police officer having jurisdiction or someone in a related wing-like intelligence does not need a permit to enter. Section 22 of the Police Act, 1861, provides that 'Police-officers [are] always on duty and may be employed in any part of district—Every police-officer shall, for all purposes in this Act contained, be considered to be always on duty'.

The provisions of Section 149 of the Code of Criminal Procedure, 1973, are also relevant. 'Police to prevent cognisable offences—Every police officer may interpose for the purpose of preventing and shall, to the best of his ability, prevent the commission of any cognisable offence.' If apprehension exists that a cognisable offence may be committed inside a particular area like a wildlife sanctuary or there is possibility of a design by someone within such area to commit a cognisable offence outside, it is the statutory duty and responsibility of the police to interpose.

In the background of established knowledge that left-wing extremists base themselves in forests and sanctuaries, and this having already happened in the past, it becomes all the more necessary for the police to gather intelligence on activities inside the sanctuary and interpose whenever required. Obstruction to such interposition would render the person so obstructing liable under section 186 IPC for 'Obstructing public servant in discharge of his public functions'. More importantly, from a plain interpretation of criminal law, if such public servant—the police here—has been obstructed from taking steps to prevent commission of such offence and the offence is consequently committed, the person who would have caused the obstruction would be liable for abetting its commission. This may be illustrated taking the example of one of the 17 cases registered in connection with the left-wing extremists' attacks inside Simlipal.

The first Naxalite attack on Chahala, Jashipur PS Case no. 24 dated 29 March 2009, was charge sheeted for trial u/s 395/ 427/ 436 / 120(b) IPC/ 25 Arms Act/ 17 Cr. LA Act/ 3 PDPP Act. The maximum sentence that each accused could be awarded by the trying court of session if convicted would be imprisonment for life u/s 395 of the Indian Penal Code, imprisonment for a second term of life u/s 436 IPC, two years u/s 427 IPC, seven years u/s 120(b) IPC, and some more u/s 25 Arms Act, 17 Criminal Law Amendment Act and Prevention of Damage to Public Property Act. Presuming that commission of this offence could have been prevented if the police had been able to do preventive reconnaissance and intelligence collection inside the sanctuary, it may be inferred that their inability to do so because of denial of entry into the area where the offence was committed facilitated the commission of the offence. Consequently, those responsible for directly or indirectly obstructing the police from taking necessary preventive action by restricting their entry into the park could be responsible for abetment of the offence committed and consequently liable u/s 109 of the Indian Penal Code to life imprisonment for abetting the offences u/s 395 and 436 IPC. This is not taking into account the other sections under which the case has been charge sheeted and the 16 other similar cases registered for which the applicability of law would be the same.

Even while things appear to have stabilised, reports that came out of Simlipal a few years back, even after militants had been flushed out, indicated that large-scale poaching has been continuing in the reserve. A feature in a credible English daily reported in August 2013, that serious staff shortage and absence of full-time officers in key areas have left the park at the mercy of poachers and smugglers. Some of the critical areas where such poachers and smugglers have entered are now reportedly running the risk of a breakdown in management.[24]

Compounding the problem as already mentioned before in this chapter, another respected English daily in September 2013, quoted scientists of the Wildlife Institute of India having said that the prey base in Simlipal had become so low that ensuring a sustainable habitat for tigers had become a problem.[25] For a park spread over a massive 2,750 square kilometres where potential prey like the proliferating chital, sambar, barking deer, and wild boar numbered some thirty thousand or more not so very long back, inability to sustain a habitat for tigers that now number just a few over 20 is ominous.

In the wake of its inability to enforce adequate protection measures, the decision of the Simlipal Tiger Reserve management to check poaching by not allowing anyone at all other than its own staff into the core area puts the clock back to the same circumstances that precipitated the calamity that hit it not very long back. If poachers have been converging on the park again in the backdrop of the severe staff crunch that the Simlipal management has self-admittedly been suffering from, it may not be long before extremists also do so again given the suspected links that poachers and militants seem to have developed. The question that arises is who will be responsible if armed militants enter the park again and what will the liability of those responsible be?

As the consequences of indirectly encouraging Naxalite terrorism to simmer in the reserve would probably not have been known to the earlier park managements, it may be argued that there had been no *mens rea* or conscious motive on their part. But the situation having happened once before, if similar circumstances are replicated for it to recur, and like poachers are doing, militants infiltrate the park a second time, the park management may not be able to claim that it was not aware of the possible consequences.

With the Wildlife (Protection) Act, 1972, clearly stipulating that public servants on duty cannot be prevented from entering sanctuaries, the wildlife authority that would have passed such orders preventing the police from entering the park, if apprehension exists of possible infiltration by armed militants, could be guilty of indirectly helping such militants further their cause and abetting the offences that they might thereafter commit. A discretionary executive order based on a statutory empowerment under law to keep a particular area out of bounds for all cannot ordinarily supersede the statutory authority granted to another public servant under law.

The seriousness of the matter can be judged from the fact that offences of dacoity and arson which the militants committed before in Simlipal are heinous enough offences punishable with life imprisonment, but waging war against the state which is the professed aim of Naxalites provides for the death penalty u/s 121 IPC, the highest punishment in the laws of the country. The gravity of the situation can be appreciated from the fact that abetment of such offence punishable with death is punishable with up to 14 years of imprisonment under Section 115 of the Indian Penal Code.

As this issue of interpretation of different overlapping laws is a matter having grave security implications, particularly in the background of a former Prime Minister of the country having described Naxalite militancy as the greatest threat to the security of India, there is an imperative need for the administrative and legal fallout to be adequately scrutinised at appropriately high levels of concerned authorities. It would be ironical if prevention of poaching crimes punishable with a sentence of up to seven years of imprisonment, extremely important though it is, should be at the expense of allowing far more serious crimes against the state punishable with life imprisonment or death.

Instead of poaching and militancy being allowed to pool their resources together as seems to have happened, there needs to be adequate and proper integration of wildlife and security management in the country or we may be missing the woods for the trees—compromising the security needs of the country in the process of containing poaching and protecting wildlife. That would be a heavy price for the nation to pay, presuming it has not already been doing so.

Notes

1. Available at odishawildlife.com
2. Report of the 2 member committee constituted by the NTCA on 3 June 2010 to 'assess situation in the Simlipal Tiger Reserve in the wake of widespread elephant poaching'; Belinda Wright, 'Assessment of Recent Elephant Poaching in Simlipal Tiger Reserve', report submitted to NTCA, 30 July 2010.
3. Research publication for PhD by B. Mohanty, 'Wildlife Poaching in Odisha', chapter on 'Mammals–Elephant' under 'Wildlife Species in Trade'.
4. H. K. Rout, 'North-East Poachers Sneak into Simlipal, Kuldiha forests—Back in 1994, Lisu Tribals Were Arrested for Mass Killings of Elephants Here', *The New (Sunday) Indian Express*, Bhubaneswar, 14 October 2012, Baripada.
5. Ibid.
6. V. R. Ramanath/*TNN*, 'Low Prey Base Worry for Simlipal', *Times of India*, Bhubaneswar, 18 September 2013.
7. Available at https://en.m.wikipedia.org/wiki/Tiger_reserves _of_India.
8. Project Tiger: Wild Life Conservation in Odisha—Odisha Wildlife, available at odishawildlife.org>project tiger.
9. Status of Tigers in India, 2014—National Tiger Conservation Authority/Project Tiger, source Government of India, MoEF & CC, available at projecttiger.nic.in.
10. *The New Indian Express*, 'Odisha Government Figures it Out, Says "We Have 40 Tigers"', 18 May 2016, Bhubaneswar.
11. M. Bhardwaj, *The New Indian Express*, Bengaluru, 29 September.
12. Status of Tigers in India, 2014—National Tiger Conservation Authority/Project Tiger, source Government of India, MoEF & CC, available at projecttiger.nic.in.
13. 'Global Afghan Opiate Trade', United Nations Office on Drugs and Crime Report by H. Demiburken, AOTP Programme Manager, and H. Azizi, AOTP, UNODC, Vienna, July 2014.
14. 'The UNODC Sahel Programme Delivers across the Region', Contribution to the United Nations Integrated Strategy for the Sahel (the UNODC Sahel Programme, 2013), available at www.unodc.org/westandcentralafrica.
15. 'Criminal Nature : The Global Security Implications of the Illegal Wildlife Trade', 20 June 2013, International Fund for Animal Welfare, available at www.ifaw.org/united-states/resource-centre/criminal-nature-global-security-implications-illegal-wildlife-tra-0.
16. Letter no. 744/AN Cell dated 13.07.2013 of Superintendent of Police, Mayurbhanj, Baripada.
17. Report of NTCA Committee set up on 14 July 2009 to 'assess situation in the Simlipal Tiger Reserve in the aftermath of the naxalite attacks', 22, available at http://projecttiger.nic.in/whtsnew/Simlipal_Appraisal_Report_FINAL [1].pdf.
18. Letter no. 744/AN Cell dated 13 July 2013 of Superintendent of Police, Mayurbhanj, Baripada.
19. Report of the 2-member committee constituted by the NTCA on 3 June 2010 to 'assess situation in the Simlipal Tiger Reserve in the wake of widespread elephant

poaching': Assessment of recent elephant poaching in Simlipal Tiger Reserve, Belinda Wright/NTCA 30 July 2010.

20. '2014 All India Tiger Estimation Result', National Tiger Conservation Authority of India, available at projecttiger.nic.in.

21. V. R. Ramanath/*TNN*. 'Efforts on to Rid Poachers from Simlipal Forest Reserve area', *Times of India*, 6 May 2013, available at http://articles.timesofindia.indiatimes.com/2013-05-06/flora-fauna/39063534_1_tiger-reserve-national-tiger-conservation-authority-ntca.

22. Letter no. 6356/3WL-85/13, 6 August 2013 of J. D. Sharma, IFS, Principal CCF (Wildlife) & Chief Wildlife Warden, Bhubaneswar, Odisha.

23. S. K. Dubey, 'Maoist Movement in India—An Overview', 6 August 2013, available at http://www.idsa.in/system/files/BG_MaoistMovement.pdf.

24. *The New Indian Express*, 'Poachers Have a Field Day as STR Reels under Staff Shortage', 5 August 2013.

25. V. R. Ramanath/*TNN*, 'Low Prey Base Big Worry for Simlipal', *Times of India*, Bhubaneswar, 18 September 2013.

5

Future of the Asian Elephant

According to Megasthenes, ambassador from the Seleucid Empire to the Mauryan court in the third century BC, Chandragupta Maurya's army of 30,000 cavalry and 6,00,000 infantry included 9,000 war elephants.[1] Elephantry in the military was first practised in India, spread to southeast Asia and then went west on to the Mediterranean, as Greek king Pyrrhus and Hannibal of Carthage were among those who engaged them.[2] As mentioned earlier, Kautilya's *Arthashastra* mentions that elephants found in the forests of present-day Odisha were best suited for war. Interestingly, Alexander's Admiral Onescritus is on record for having said that the elephants of Taprobane, known as Sinhala thereafter and Ceylon later, were 'bigger, more fierce, and furious for war service than those of India's', a statement corroborated by Megasthenes. Greek writer Aelian mentions export of Sinhala elephants from around 200 BC to Kalinga—the place in India where the best elephants came from.[3]

During the reign of Sinhalese kings, it was a major crime in Sri Lanka to kill elephants, the punishment for which was death. Elephants were thereby well protected. With the onset of colonialism, however, elephants came to be declared as an agricultural vermin and there was a bounty for those who killed them. Needless to say, tuskers were killed in great numbers for sport. Unlike India, that had vast forests for elephants to retreat to, Sri Lanka did not give them the chance and they were decimated to the extent that tuskers were more or less completely liquidated. Today, in the land which produced the best male war elephants, the vast majority of whom bore tusks, the percentage of ivory-bearing elephants in the male elephant population is just 7 per cent, and in the total elephant population, an insignificant 2 per cent.[4] Ironically, it is because male elephants in Sri Lanka today are almost entirely tuskless *makhnas* that poaching for ivory has ceased to be an issue of concern.

There is an important lesson here. Of the three sub-species of the Asian elephant, it is the Sri Lankan, aptly named *Elephas maximus maximus*, that is the largest. The Sri Lankan elephants of today are, however, believed to be smaller than those that are depicted in pictures to have been exported back in 200 BC, or for that matter those photographed in the nineteenth century during the days of colonialism. Both, the diminished size and absence of tusks, in the largest sub-species of the Asian elephant are probably because the best physical specimens were over a period of time eliminated to the extent that only those of lesser pedigree were left behind to breed and pass on their genes to subsequent generations.[5] From a little under 20,000 elephants at the beginning of the twentieth century, it is believed that there are about 4,000 left in the country now, a drop of 80 per cent in the space of a hundred years or so.[6]

The Sumatran elephant, the other sub-species of the Asian elephant and the smallest of the three, has, like the elephant in Sri Lanka from where it got its name, been confined largely to the Indonesian island of Sumatra. Simultaneously, with intense deforestation and death related to conflict, the Sumatran elephant despite smaller tusks that males have, has also been targeted for ivory poaching such that more than half of Sumatran elephants disappeared in just about 20 odd years. It is believed that there are less than 2,000 of them left, and while poaching continues to feed ivory demands outside the country, poisoning of elephants when they stray into agricultural areas also poses a serious threat to the longevity of the sub-species.[7]

Unlike its Sri Lankan and Sumatran cousins, who are located almost entirely where they get their names from, the Indian elephant is not confined just to India but to mainland Asia that includes Bangladesh, Bhutan, Cambodia, China, Laos, the Malay Peninsula, Myanmar, Nepal, Thailand, and Vietnam. However, except for Myanmar that has some 4,000–5,000 elephants, and Thailand and Malaysia between 2,000–3,000 each, other countries have them only in hundreds.

India is home to about 60 per cent of all Indian elephants, numbering between 27,000 and 28,000 in the country.[8] Even while elephants in India have had the benefit of larger land area available, the burgeoning human population and increasing loss of forests to agriculture and mining operations are negating the advantage it once had. With the Sri Lankan elephant having gone the way it has and the Sumatran going where it is, the circumstances that have caused them are being replicated in India and should sound warning bells lest the Indian elephant, already under threat, goes the same way.

MIKE, a programme started under the Convention on International Trade in Endangered Species (CITES) in 2003 to monitor the illegal killing of elephants in Africa and in South Asia, was constituted to provide information for appropriate management and enforcement decisions in elephant range states. Among its main objectives were to measure levels and trends in the illegal hunting of elephants and to determine the factors causing or associated with such trends for corrective action to be taken.[9] In this regard, it would be relevant to examine a few of the recent forms and trends in elephant killing in India vis-a-vis the main thrust of this study.

Electrocution Deaths, Train Hits, and Poisoning

Some new forms of death have been visiting the elephant in recent times and adding to the considerable pressure on its well-being. Even while the electrocution death problem has become widespread in various states, Odisha is now among the states which has the highest in the country.

The electricity distribution network commences with 33 kilovolts high voltage cables from power houses branching into 11 kilovolt lines which lead to lower voltage transformers and connections to consumers. With the objective of providing electricity to all villages in all parts of the state, long stretches of line of 33 kilovolt and 11 kilovolts run through forest areas. For a variety of reasons, like not being tightened enough at the ends or the base structures of transmission towers not being strong enough, cables often sag low. As they carry high voltage power, they are deadly when living creatures come in contact with them. Even insulation is sometimes not quite adequate, and rains, over a period of time, weaken the concrete base that holds up the electrical tower installations; there are instances where elephants had rubbed their itchy bodies and rumped against the metal girders at the base of the tower—causing them to bend.

With elephants being taller than other animals, the chances of them accidentally touching such live lines is naturally high. Even while those in areas familiar to them do accidentally come in contact with power cables that might have sagged, noisy and violent dynamite blasting in mining operations in forests disorientates them into straying into new areas where they are more likely to come in inadvertent contact. Accidental electrocution deaths are particularly high in October maybe because after the rains overgrown tree branches and shrubbery conceal further disseminate the effect of the live lines. Being solitary animals, tuskers often get electrocuted singly while more than one usually dies when it is a female or juvenile. Herds tends to try and come to the rescue of fellow members in distress and often get electrocuted themselves.

While accidental electrocutions keep happening, farmers often lay out live wires to protect their crops from marauding herbivores, and poachers set traps with similar live electric cables to target tuskers, not all their victims actually being tuskers. Of the 142 electrocution deaths in the state from 2000–01 to 2015–16, 74 have been accidental electrocutions and as many as 68 deliberate.

Given the increasing frequency of such deaths, a four-member committee constituted by Government of India's Ministry of Environment and Forest in 2010 examined the high rate of electrocution deaths in the state and suggested remedial measures.[10] However, when there were 16 deaths from accidental electrocution and as many as 28 by deliberate electrocution in just three years thereafter, there were some important issues raised.

Even while the wildlife authorities and the energy department traded charges of responsibility for electrocution incidents, the mutual recriminations seemed to be getting priority than the need of trying to stop the killings. Several write-ups in major English and vernacular dailies mentioned that senior wildlife authorities had conceded that poachers were deliberately electrocuting elephants through laid-out live wires but were blaming the power distribution companies for not installing adequate safeguards. Conversely, the energy authorities had questioned why wildlife authorities were not identifying the elephant killers and taking action under law as empowered.

The wildlife authorities had registered 5 cases for 52 incidents of deliberate electrocution until 2012–13.[11] In the 3 years between 2013–14 and 2015–16, 8 cases were registered for 16 deliberate electrocution killings and 23 people were arrested. There have been a total of 142 deaths by electrocution since the beginning of this century at an average of a little under 9 a year. However, 14 electrocution deaths were reported in the year 2015–16 alone of which 9 were deliberate killings.[12] This would suggest that the threat to the elephant on this front may only be increasing.

While there had been the occasional death of an elephant in the past because of a train hit, the spate of deaths that has happened since then has gained considerable notoriety in recent times. Even as the forest cover and elephant habitat have been shrinking, considerable expanses of ore-rich forest areas are being explored for mining. To facilitate movement of the excavated ore to ports and other destination points, new rail links have been laid. Many such rail lines have been extended in dense forests that have traditionally been elephant movement routes. There have been several occasions of trains mowing down elephants crossing the tracks particularly at night time when the light and noise of the oncoming locomotive causes them to lose orientation.

This has been quite extensive in states like West Bengal where 35 elephants have died in train hits in 5 years and Assam where 21 died. Odisha too has not been too far behind with 16 deaths, 14 in the last 4 years of which 11 were in 2012–13 alone.[13] A war of words between the wildlife and the railway authorities has taken off each time when there is an incident. The former insisting that trains move slowly in the entire forested area where elephants are ordinarily found and the latter seeking specific information on elephant movement because slowing down trains completely would adversely affect the railways' finances.

Even as the issue drew the attention of the Supreme Court of India in the form of a public interest litigation, the Ministry of Environment and Forest and the Ministry of Railways after a series of deliberations came up with a joint advisory that railway authorities restrict train speed in key elephant areas so as to reduce braking distance, use whistle at curves to warn elephants that might be on or near the rails, and for wildlife authorities to ensure tracking of elephant movement so that train drivers can be forewarned about specific elephant presence.[14]

This may have had some deterrent effect, because after 25 deaths in 2012–13 and 17 in 2013–14, there have been 6 train-hit deaths in the country in 2014–15. Odisha that had witnessed 11 deaths from train hits in 2012–13 has in the years since then seen only two more. But here again, the question arises whether inter-departmental or inter-ministry coordination issues have to go all the way up to the Supreme Court to be decided when they could and should have been settled at the administrative level itself?

Of the states in respect of which particulars of elephant deaths by poisoning are available with Project Elephant of Government of India, Odisha has had by far the highest share with 18 deaths in 5 years. Poisoning has been a major cause of decline in the population of the Sumatran elephant, as most of them being instances of farmers trying to keep them off their crops in course of amongst the fastest deforestations anywhere in the world. In Odisha, however, it is primarily tuskers that have been targeted because many of the poisoning cases happened in the Simlipal Tiger Reserve and neighbouring areas. Out of 18 elephants 16 were killed in the 3 years between 2010–11 and 2012–13 when poaching in the reserve had become rampant in the aftermath of the Naxalite attacks. However, not many of those actually killed were tuskers, more of females and juveniles getting to the poisoned baits earlier and being victims of collateral killings. Death casued by poisoning has also declined, and there were two casualties in the last three years.

Elephant Distribution, Poaching, and Militancy

The continent of Africa stretches across 30 million square kilometres of mostly dry savannah grassland and is about 10 times the size of India that predominantly has tropical forests. The animals differ in numbers, variety, and density. Animals like the lion and the elephant are common to both places, but the African lion differs from the Asiatic variety and the African bush and forest elephants are different from the Asian and the Indian. Compared to the 4,00,000–4,50,000 African elephants, both males and females having tusks, there are around 27,000 in India where it is only the male that has them.

Despite the differences, however, there are similarities in circumstances and environment particularly relating to the behaviour of herbivores. Even while the anticipation of rain and resultant fresh grass that induces the annual migration of two million wildebeest and zebra between Serengeti in Tanzania and Maasai Mara in Kenya is the biggest seen, the need to drink causes elephants in densely elephant-populated water scarce countries like Botswana too to move across borders much like elephants in India do from one state to another for feeding. With regard to elephant movement and behaviour, boundaries between countries in Africa are about as porous as are those between states in India.

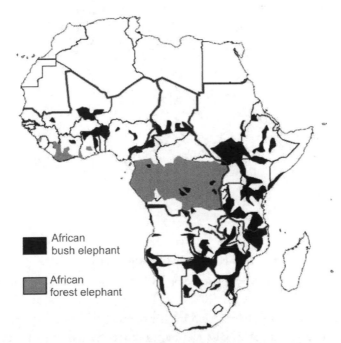

Figure 25 Distribution of elephants in Africa (4,00,000–4,50,000)[15]

There is a difference though. Unlike Africa, where elephants are distributed in all the countries in the continent except the north, in India they are largely confined to four regions. These patches of elephant presence in the north-eastern, central-eastern, north-western, and southern parts of the country, even though some of them might have been more contiguous in the past, are at considerable distances from each other today. Consequently, even while they ordinarily remain residents of the states they are in, elephants do move from one state to another seasonally within their respective regions for feeding. The Indian wild elephant population distribution could be pictorially depicted as below.

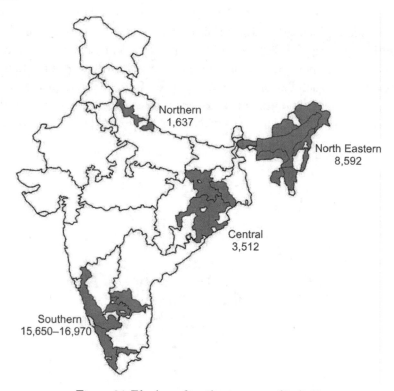

Figure 26 Elephant distribution map of India[16]

Note: Map not to scale and does not represent authentic international boundaries.

It would be desirable to mention in this context that as per the census done in 2017, there are about 27,312 elephants currently living in India. This is perhaps a more accurate figure than earlier estimation because, unlike before, the counting was done simultaneously in the states of each region. Elephant population in the country is believed to have more or less stabilised around that figure but the primary concern is the number of tuskers living in earlier years and the number of tuskers currently residing in India.[17] Pending official release of the final census figures of 2017 however, it would be appropriate to keep the previous estimate to be the base figure.

In case of the African elephant, its population is spread across the entire continent except the north and its habitat is largely contiguous. The density of elephant presence is relatively less in the western and eastern parts of the continent, comparatively more in the central African region, and is the maximum in the southern countries as well as a few eastern ones like Kenya and Tanzania that have traditionally been wildlife powerhouses.

Assuming that the forest elephants in Central Africa might be threatened with extinction given the rate at which they are being killed, it is significant that almost all the central African countries like Chad, Cameroon, Central African Republic, South Sudan, and Democratic Republic of Congo have been in the grip of severe insurgency and civil war for some time now. Not insignificantly, these have coincided with the massacre of elephants in recent years in central Africa.

Even the United Nations Office on Drugs and Crime (UNODC) reports that contraband heroin from Afghanistan has infiltrated African countries in a big way and has been financing criminal operations, better coordination between narcotics enforcement agencies of affected countries at its behest has led to militant groups being pressurised into moving elsewhere for their finances. Without much overhead costs or necessary infrastructural requirements, mass killing of animals, particularly elephants and rhinos, has become easier for bringing in the finances.

Elephant ivory and rhino horn are most sought after because of the huge prices they command. Terrorist groups like Al-Shabab in Somalia have been foraying into Kenya for slaughtering elephants for ivory in the east while Cameroon in the west recently saw probably the worst-ever single massacre of elephants. The bottom line is that central Africa has lost over 60 per cent of its elephant population to ivory traders in just a few years and, if continued to be killed at this rate, there will be an existential crisis. Indeed, more elephants are believed to be dying than are born each year.[18]

Importantly, while it was generally believed that it is the forest sub-species of the African elephant that is being targeted, there are reasons to believe that the Savannah or bush elephants are not safe either. According to the Great Elephant Census announced in 2016, a decline of 30 per cent, numbering around 1,44,000 bush elephants, was found to have occurred in 15 of the 18 countries where the study was taken up. It is believed with 3,52,271 bush elephants left in the wild in 18 countries today, the rate of decline was found to have been 8 per cent every year over the previous eight. The one saving grace is that China, a major destination for ivory, is believed to be moving towards closure of its ivory markets by the end of 2017.[19]

In India, the predominant armed militancy movement has been of left-wing extremism that has been at it for close to two decades now, at one time covering one third of India's districts and its landmass. The manifest of these Maoist militants explicitly speaks of the creation of a red corridor of their dominance separating India into east and west. The proposed corridor, consisting of the states of Bihar, West Bengal, Jharkhand, Odisha, Chhattisgarh, Andhra Pradesh, Telengana, parts of Karnataka, and Maharashtra, has Nepal on the north, adjacent to Bihar and West Bengal. This Himalayan country has pronounced Maoist 'red' influence and is a nodal point of the illegal wildlife trade market, mostly to China.

Separatist insurgency has also been seen in good measure in the north-eastern states of India as well as in the northern state of Jammu and Kashmir even though the latter is not directly relevant to the scope of this study. In the north-eastern states elephants are being killed for ivory as much as the one-horned Indian rhino is for its horn, feeding the Chinese black market through the porous borders of Manipur and its sister states with Myanmar. With poachers in the north-east using the latest sophisticated firearms, it does not need special investigative abilities to infer that poaching and militancy have joined hands. Indeed, it has been officially acknowledged by investigating authorities that poachers are being provided AK-56 and AK-47 automatic assault rifles by militant separatists and that poaching helps fund north-east India rebel groups.[20]

Among the large number of rhinos poached in 2016 was one that was shot during the visit of the Duke and Duchess of Cambridge to Kaziranga National Park, not far from where they were put up.[21] Such incidents suggest that the militant–poacher combine is getting increasingly brazen. Population of elephants in the north-east, particularly tuskers, has also dropped quite substantially over the years, which is testimony to their being increasingly targeted. The illegal trade from north-east India is not just in respect of elephant ivory and rhino horn but also for other living and dead wildlife products.

In the south where there is the largest concentration of the Indian elephant comprising half the numbers now in India, the three states of Tamil Nadu, Karnataka, and Kerala where they are primarily located have been in the grip of an elephant poaching frenzy not seen since the days of the notorious brigand Veerappan who killed 75 tuskers in just a decade. Even while 375 elephants had been killed in these states between the mid-1970s and 1990s, 100 tuskers are believed to have been killed in just two years presently.[22]

The map below shows militancy-affected areas of India. A comparison of the map in Figure 26 showing elephant distribution in the country with that in Figure 27 depicting areas affected by insurgency and armed militancy is revealing. As may be seen, just about all the areas where elephants are found in the country are those where there is armed militancy as well.

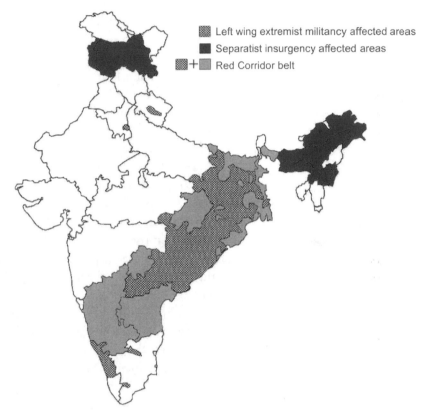

Figure 27 Militancy-affected areas of India[23]

Note: Map not to scale and does not represent authentic international boundaries.

It would be erroneous to compare the African elephant poaching situation with the Indian merely on the basis of statistics. Both males and females of the African variety bear tusks so poachers do not discriminate between them and have often machine-gunned entire herds. Except for juveniles that do not yield much, every single elephant killed provides ivory to the killers. In the case of Indian elephants, it is not just that only the male carries tusks but these males are solitary animals that are singled out and hunted. Consequently, the numbers killed would obviously be a lot less than those that are killed in Africa, but given that only 27,000 or so elephants are left in India and the fact that the the number of tuskers would be less than 10,000, those killed make a serious dent, not just to the gender ratio but to the proportion of tusk-bearing males as well.

As may be seen from a comparative study of the maps showing distribution of elephants and presence of militancy in India, the north-western habitat does not have too many elephants but poaching does exist there as does left-wing militancy in some areas.

Likewise, the southern habitat, where concentration of elephants is the highest in the country, also comprises a substantial portion of the red corridor of left-wing extremists and has had considerable militant activity in and around the area. The fact that poaching of tusker elephants in these southern forests has escalated sharply in recent times cannot be dissociated from the presence of armed militancy there. As has been mentioned earlier in the work, two organised illegal operations from the same forests, even if they might have started separately, cannot remain disconnected for long given the clandestine nature and physical proximity of such operations and the fact that their common adversary is the authority of the state. Between the 1970s and 1990s, it was one brigand leader's network that had decimated tusker elephants in the south. Presently, there are grounds to believe that it could well be the militant–poacher combine that has caused its resurgence.

The remaining two elephant zones of the country are heavily affected by insurgency. While just about every state in the north-east is seriously hit by separatist insurgency and continues to witness widespread violence and poaching, the states of the east-central zone, namely Odisha, Jharkhand, West Bengal, and Bihar have been affected by militancy and poaching for many years now.

As China is the major destination in this illegal wildlife trade, Myanmar in the north-east and Nepal in the north serve as conduit states. As the north-eastern border is porous as it is, movement of wildlife goods into Myanmar with the joint connivance of organised poachers and separatist militants is rampant. Likewise, as the states of Odisha, Jharkhand, West Bengal, and Bihar are heavily Maoist-affected, wildlife trade from these areas into Nepal where Maoists have a major influence is easily facilitated. Not surprisingly, a substantial inflow of arms for militants comes from Nepal—which is immediately north of Bihar—and from across the borders in the north-east, the very same areas that the wildlife trade is routed through beyond the country.

Even while deliberating on the fate of elephant, it becomes necessary to focus on the two other heritage animals of India that have become victims in this high-return illegal trade. The Indian one-horned rhino that is found almost entirely in the north-east is being massacred for its horn which fetches enormous prices in the international black market. So is the tiger, whose body parts are used for a host of traditional medicines in China from cure of arthritic joints to treatment of impotency. With increasing consciousness about the menace of the illegal wild animal parts trade, the World Wildlife Fund recently came with a move to shut down tiger farms in Asia. Reports indicate that there still are two hundred tiger farms across China, Laos, Vietnam, and Thailand. The recent discovery of a large number of tiger cub carcasses in a freezer in Thailand's Temple of Tigers precipitates a swift reaction, not only within the country where they were shut down but also internationally.[24]

In recent times, there have been a lot of deliberations on the investigation of wildlife crimes and the use of latest forensic science applications including DNA fingerprinting to solve poaching crimes. The international police body Interpol too has been playing a proactive role in this regard as are the investigative agencies of various countries including the Central Bureau of Investigation in India. This is a welcome development in the fight against wildlife crime. However, there would be better effectiveness if measures to optimise prevention were also intensified.

Apart from the tusks of elephants and the horn of rhinos, the remaining body parts of the animal is largely worthless, and after the tusks have been extracted and the horn sawn off, the body is left abandoned. As both the elephant and the rhino are large animals, their carcasses are usually discovered, sometimes early sometimes a little late. Investigation can commence with such discovery and the latest scientific tools put to good use. But there is a difference in respect of the tiger whose skin, skull, bones, canines, claws, whiskers, and even its penis fetch huge prices and are in great demand. With all body parts removed and only the flesh of the animal left behind, which in any case is promptly consumed by scavengers, there is very little of the tiger left to begin investigation with. Indeed, the killing of tigers often escapes attention except when there is a seizure of skin or bones which is when it comes to notice.

Consequently, unlike the elephant or the rhino, an assessment of the number of tigers actually killed may not be very accurate. This probably explains why Project Tiger was hailed as a huge success story until the National

Tiger Conservation Authority (NTCA) announced one not so fine morning that there were just a few over 1,400 left in India's wilds. Hence, even while investigation is very important, there needs to be greater focus on steps to prevent killing in the first place because as often is in the case of the tiger, investigation cannot commence unless there is something to begin it with. Prevention is essential because investigation commences only after the animal has been killed and each time that happens it means that there is one less animal around even if the person responsible has been subsequently caught.

When the NTCA announced in 2014 that the number of tigers in India had risen to 2,226, it was greeted with euphoria because it was for the first time ever that an increase in the tiger population was reported. However, there have been disturbing reports since then suggesting that the poaching mafia is actively catching up.

If figures provided by the Wildlife Protection Society of India are to be relied on, as many as 50 tigers were killed by poachers in 2016 out of a total of 132 that died. Until July 2017, 78 tigers have already died of which at least 23 have been confirmed cases of poaching. A particularly worrying factor is that the 15 states where tigers have been killed or body parts seized during the past 3 years have included Arunachal Pradesh, Assam, Nagaland, Bihar, West Bengal, Odisha, Chhattisgarh, Tamil Nadu, Karnataka, Kerala, Maharashtra, Madhya Pradesh, Uttarakhand, and Uttar Pradesh which have all seen some form of armed militancy. Investigations have conclusively established in at least the few cases where leads were followed up that the skin and body parts of the tigers had gone to Kathmandu, almost certainly on their way to China.[25]

With the demand for wildlife items being almost insatiable because of the huge industry that they sustain and the enormous prices that they fetch, there will always be suppliers who will come up to replace those that are put away. So a day might be reached when such individual killings add up to a stage when it might just be too little too late. Which is why there should be a greater focus on preventive measures and some attitudinal change to facilitate better synergy between stakeholders in this losing game.

Attitudinal Insularity—Need for Change

In the aftermath of the Naxalite attack in Simlipal that was accompanied by unabated poaching—sanctuary authorities admitted that distribution of animals in the reserve had fallen to just four per square kilometres[26]—wildlife authorities of Odisha alleged at high-level government meetings that since the police had failed to provide them security, it had become difficult for them to prevent poaching in the reserve. With regard to increasing number of elephant deaths by electrocution, a frequently heard complaint was that the energy authorities were not ensuring adequate safeguards for which the wildlife authorities were unable to do much. When there were increasing instances of trains mowing down elephants, a statement often issued was that the railway authorities were not heeding their request to cut down train speed and the wildlife management was therefore not in a position to prevent such gory elephant deaths. The common refrain in the interface with the three other organisations concerned was that these organisations were held responsible for the wildlife department's inability to cope with the poaching problem and other preventable elephant deaths.

To examine the three separately, difficulties faced by the police over the years in intelligence collection on activities inside sanctuaries have been elaborated earlier. Starting with Simlipal, the fact that every tiger reserve in the state and just about every wildlife sanctuary has become in due course a hideout or passageway for militants cannot be a matter of mere coincidence. After restricting entry of the police into sanctuaries on the grounds of preventing disturbance to wildlife and then charging the same police with failing to provide security in the sanctuary when militants take over does not quite reflect the best synergy between the enforcement agencies concerned, both of whom are stake holders. At the same time, there are genuine complaints too about the police not according much priority to wildlife-related offences, ostensibly because they already have their own crowded charter of duties to perform.

In the matter of electrocution deaths, not only is the wildlife authority competent to take action under law against poachers deliberately electrocuting elephants but also is its duty to do so. Even while blaming the energy authorities for in action, records show that wildlife authorities in Odisha had registered a mere five cases under the Wildlife (Protection) Act, 1972 for deliberate electrocution of elephants when the forest minister of Odisha himself, in a statement on the floor of the State Legislative Assembly on 27 August 2013, stated that 53 elephants had been killed by deliberate electrocution until then.[27] Since then, however, there has been greater alacrity, and in three years thereafter, 23 poachers have been arrested in eight cases of deliberate electrocution involving killing of 16 elephants.

With regard to train accidents too, railway authorities have largely been blamed for not controlling the speed of trains in forest areas. This may well be a genuine reason but the observation of a senior official of the Indian Railway Board who the author interacted with is revealing. The wildlife authorities reportedly did not accept the request of the railway authorities that wildlife personnel should track elephant movement and keep them informed so that train drivers could be alerted in time to slow down. On the contrary, wildlife authorities reportedly insisted that all train movement needed to slow down because it was not possible for them to monitor elephant movement over such vast forests.[28] The Official clarified that if all train movement in the ore-transportation forest areas is slowed down, Indian Railways could well become financially crippled because freight movement is its major source of revenue, and defaulting in timely delivery would have severely adverse financial consequences. It finally needed the intervention of the Supreme Court of India, and a joint advisory by the Ministry of Railways and Ministry of Forest and Environment at the highest level of the Government of India to arrange a meeting point.

There is an important lesson to be learnt here. If effective steps for conservation can be taken only after other departments, who have their own priorities, provide the required working environment for the wildlife authorities at the expense of their own, the tusker and the tiger that are already losing out in the battle for survival could well be doomed by the time this actually happens. Poaching and loss of wildlife are a matter that seriously concerns every right-thinking citizen, but the fact is that every government department and establishment has its own raison-d'etre for existence and its own priorities which may not always be perfectly in sync with those required by the conservation authorities. It is there where the differences must be sorted out: the lower the administrative level where this is done, the better it is for the animals concerned.

It is, therefore, imperative that there is constructive dovetailing of efforts and goals through proper synergism between the various stakeholders. It will be the heritage animals gracing our land that will ultimately lose out from this win–lose situation if it is not mended to one of the mutual compatibility and accommodation.

Another issue that needs deliberation in the background of reports of inadequate resources for wildlife protection is with regard to proposals that various committees and think tanks have gone into and recommended. Like in most establishments, these relate primarily to creation of staff for enforcement, establishment of a wildlife intelligence apparatus, procurement of requisite arms and equipment, strengthening the provisions of law, and so on. Such proposals are generally dependent on allocation of funds or are slow legislative processes, usually both.

A country like India that promotes the idea of welfare state does not have unlimited resources and those that are available have to be spread over various requirements. While wildlife management is a matter high in that list, there is a vast spectrum from the basic subsistence needs of the population living below poverty line at the state level to the need for troops to man India's untenanted borders with belligerent neighbours at the national level that merit urgent attention too.

Complete measures for wildlife conservation, as are mooted, can therefore happen gradually, and, even after sanction, actual field effectiveness after recruitment, training, procurement of equipment, and proficiency in handling them would take time. Until then, it is vital that optimal utilisation of existing resources is done in coordination with other agencies concerned rather than at their expense or the tusker and the tiger will have irrevocably disappeared. As the saying goes, if we wait till we are ready, we will be waiting all our lives. The twist in the tale here is that if we actually do wait, only the heritage animals of our country will be awaiting the end of their numbered days.

Some Findings of the Study

1. Tusker and tiger poaching is getting more widespread and is a part of a larger network of organised crime operating at the national and international levels. The rhino too is vulnerable in those areas it is found in, mostly north-eastern India.

2. Of the 21 districts affected by left-wing insurgency in Odisha and the 18 where elephant poaching exists, 15 are common. Since poachers would ordinarily not operate freely in forests where armed militants hold sway, links between the two are likely. The same is true in the north-eastern states where separatist insurgency is deep rooted as is their link with poachers.

3. Illegal ivory trade routes from central India to Nepal are through Maoist-affected districts of Odisha, Jharkhand, West Bengal, and Bihar. In the north-east, they are through Myanmar, the destination in both cases being China. Much like ultras in the north-east joining hands with poachers, organised poaching in other affected states happening with the patronage of left-wing militants is a distinct probability.

4. With only country-made guns being seized from poachers in Odisha even while wildlife authorities acknowledge that they have sophisticated weapons, there probably is a face of poaching that has not fully emerged yet and is likely to do so in the future. The use of AK-47 and AK-56 assault rifles by poachers in the north-east has been more conclusively established.

5. The widespread Naxalite attacks in the Simlipal Tiger Reserve of Odisha in 2009 suggest that militants had intimate knowledge of the park layout, particularly its core area. This could have happened only from prolonged presence and acquaintance with the reserve. The wave of poaching there further testifies to links between organised poaching and militancy.

6. Restriction of access of government officials other than forest personnel into core areas of sanctuaries may or may not help curb poaching with the wildlife enforcement machinery often hampered by inadequate manpower. But there is a strong reason to believe that such restrictions, particularly on the police, encourage and abet the larger problem of militancy and its integration with organised poaching.

7. Naxalites are known to operate from forests and sanctuaries, and if restrictions on police entry into such sanctuaries hamper intelligence collection and timely preventive action, the concerned authority so restricting could be liable for abetting militancy.

8. Almost the entire elephant inhabited regions of the country—where tigers and rhinos reside as well—are affected by insurgency or some form of armed militancy. With poaching rampant in these forests, the poaching–militancy link is in all likelihood a pan-Indian phenomenon.

9. While checking poaching is a high priority concern warranting firm measures, making it a one-dimensional enforcement process at the expense of national security would be missing the woods for the trees.

10. Narco-terrorism has been the major force behind global terrorist activities for quite some time now but the new player in the field is *eco-terrorism* which has come to stay. Unlike narcotic plants that can be grown as long as land is available, the stakes are higher in eco-terrorism because the world may ultimately run out of the animals that are essential to this trade.

Suggestions

1. Circumstantial evidence suggests that organised poaching and left-wing insurgency arc linked. This has already been established through direct evidence in the north-eastern states. There is a need for a more detailed research to unearth direct evidence of the same in areas where we are in the realm of probability.
2. Given the security ramifications that probable links between organised poaching and militancy have across the country where Naxalites are a comparative study of similar situations in other states where they exist needs to be undertaken.
3. Naxalites are known to have operated from various national parks and sanctuaries around the country, as the wildlife and particularly the tiger population have taken a hit as a result. As studies have established that Naxalites mostly occupy forests, the various parks and sanctuaries they have operated from and circumstances thereof need to be studied with reference to the impact on both wildlife and state security and remedial measures if any taken.
4. Preservation of heritage animals of the country is far too important a matter to be left to wildlife conservationists alone. Given the links with insurgency and the need to have a consolidated approach in addressing it, there is a need for an integrated approach which could begin with security enforcement officials knowledgeable in wildlife being involved in the conservation process. Unless this is done, the enforcement agencies could well continue to be working at cross purposes.
5. Tanks of appropriately knowledgeable people from amongst various stakeholders from the district administration level upwards. They need to be adequately empowered, and the lower the stage where matters are settled, the better. These could scrutinise the administrative and legal aspects of the disconnect in the fight against armed militancy on the one hand and organised poaching on the other so that the state does not lose out in the long-term stakes.

Summing Up

Huge stockpiles of ivory accumulated over the years in many African countries have been revenue earners in one off disposals. The difficulty is that there is no way of discriminating between such ivory and that taken off recently poached elephants which keeps feeding the market. Kenya has taken the lead in not just setting ablaze the biggest ever stockpile of elephant tusks but putting an end to game hunting. However, many countries where the concentration and population of elephants is high, particularly the ones in southern Africa, continue to allow sport hunting of elephants. Among them are Botswana and Tanzania where elephants number in excess of a hundred thousand, and given the limited land available for agriculture, they find it as a way to keep their numbers in check. The ivory taken off from both males and females feeds the market as indeed the coffers of the countries that allow them under license.

In India, there is no luxury of plenty. With militancy affecting public order in large tracts of the country, the source of its funding has largely been through narcotics trading, intimidation, and extortion. In many left-wing affected states, however, thousands of acres of cannabis cultivation that have been used to provide for the ultras keep getting located and destroyed. With the enforcement apparatus also coming down on drug dealing, this source of funding for insurgents has gradually been dwindling. Much as it has been in central African countries, the less risky but highly lucrative option of wildlife trade has offered an alternative option. Ironically, our inability to synergise the operational goals of various enforcement agencies has only facilitated the collusion between insurgency and poaching.

Plate 25 From the suave aristocrat

Plate 26 To the dashing debonair

Plate 27 And the senior statesman

Plate 28 To junior itinerants

Plates 29 and 30 Tusks of the extinct woolly mammoth

The mammoth which appeared on earth five million years back became extinct some 10,000 years ago. Many of those that died lie buried in countries that have now been allowing them to be exhumed and their tusks to be exported. Plates 29 and 30 photographed by the author, show tusks of the extinct woolly mammoth imported from Russia by an antique crafts dealer in upmarket Hong Kong who has them intricately carved by a team of artisans and sold to a special clientele.

He explained that all the displays in his showroom were of ivory taken off mammoths that had lain buried for tens of centuries in the icy plains of Siberia. To cater to the taste of his clients, who pay huge prices for them, the tusk is carved as desired by the customer but some portion of it is left untouched to give that antique feel to it. Significantly, the tusks that he now imports are 10 to 12 feet long, smaller than the 14 to 15 feet ones that were in circulation some years back but are no longer available.

The fate of the living Asian elephant today is not dissimilar. Those that have had the largest tusks are the ones that have been targeted first and despite the few that still have them now most that carried big ivory on them are now gone.

Otto von Bismarck had once famously said that it is wise to learn from the experience of others. In respect to the Sri Lankan elephant, tuskers were systematically brought down such that only a miniscule lot of males are left with tusks now, not enough to pass on their genes to make a difference to progeny. Ironically, this place where the biggest Asian elephants came from, where 90 per cent of males once had tusks, no longer faces the problem of poaching because those left will not make much of a difference in terms of volume of ivory.

Sumatran elephants still have male tuskers about them but there are so few left that they are already on the brink of extinction and on the critically endangered red list of the International Union for Conservation of Nature (IUCN).

We do not have to look beyond our neighbourhood Sri Lanka to know what can happen to our tusk-bearing elephants if we do not get our act together now when it matters. In the context of the Asian elephant, more singularly the Indian elephant, tusks mean vulnerability. It is the Indian elephant that still has a reasonable proportion of ivory-carrying males but the rate at which they are being liquidated rings more than alarm bells. Maybe, many centuries down the line if we have not been able to keep ivory-carrying elephants from extinction, some of them that might have died otherwise and remained buried might be exhumed one day and the ivory on them traded the way mammoth tusks are today.

Links between militants and poachers in Africa have been confirmed and it has been conclusively established that terrorist activities are funded in considerable measure by the illegal ivory trade.[29] In respect of the Asian, particularly the Indian elephant as also the tiger and the rhino, there is direct evidence of links between insurgents and poachers in the north-east and strong circumstantial evidence of the same in central India, possibly in the south too.

The only means of tackling the joint menace of wildlife poaching and armed militancy is to have some degree of integration between wildlife enforcement and the security enforcement apparatus. Each going its own way has obviously facilitated poaching and insurgency getting together. For this, it is important that security officials with knowledge of wildlife are brought into the loop of wildlife management so that decisions that might inadvertently be to the detriment of either can be avoided. The heritage animals of India, the elephant, the tiger, and the rhino, owe it to us for their survival as species.

With time running out, it is important that there is adequate cohesion of purpose and utilisation of available resources so that this multidimensional problem can be tackled on its different fronts. As was said by the spokesperson of the International Fund for Animal Welfare, 'We can't see this as an environmental problem anymore, when it has grown into a criminal and security one.'

Significantly, the international body monitoring wildlife conservation thinks alike. As aptly put by Richard Jenkins, manager of the IUCN's Global Species Programme, 'We have to look for greater cooperation with agencies that have more experience with criminal and security issues.'[30]

With the Sri Lankan elephant effectively de-tusked and the Sumatran almost at the point of no return, that is the call of the day. The Indian elephant and its bull tusker, the last of the Asian species that still has a chance, may not get another. It is well and truly under siege.

Notes

1. Available at https://en.m.wikipedia.org/wiki/ilitary_history_of-India.
2. Available at https://en.m.wikipedia.org>wiki>War_elephant.
3. 'Elephants in Sri Lanka History and Culture' by J. Jayewardene, Managing Trustee, Biodiversity and Elephant Conservation Trust. Available at artsrilanka.org; Aelian has been quoted by E. Tennent in 1859 regarding export to Kailnga from about 200 BC. Sinhala elephants' qualities were known to the Greeks as far back as the third century BC in the time of Alexander the Great.
4. Available at https://en.m.wikipedia.org/wiki/Sri_Lankan_elephant.
5. Ibid.
6. *Eleaid*, 'Elephants in Sri Lanka—Sri Lankan Elephant Population Figures', available at http://www.eleaid.com/country-profiles/elephants-sri-lanka/.
7. Available at https://en.m.wikipedia.org>wiki>Sumatran_elephant; A. Gunawan, 'Sumatran Elephants at High Risk of Poisoning, Poaching', *The Jakarta Post*, 20 April 2016, available at www.thejakartapost.com/news/2016/04/20/sumatran-elephants-high-risk-poisoning.html.

8. Available at https://en.m.wikipedia.org/wiki/Indian_elephant.

9. Available at https://cites.org/eng/prog/mike/index.php.

10. Report of the four member committee for enquiry into the cause of death of elephants due to electrocution in Orissa, 15 December 2010, available at www.scribd.com/doc/48778724/Report-of-the-4-Member-Committee-of-the-MoEF-that-probed-elephant-electrocution-deaths-in-Orissa-in-2010.

11. *PTI,* 'Elephant Deaths Rampant in Orissa', *The Hindu,* Bhubaneswar, 31 October 2012; '49 Elephants Hunted with Electric Shocks—300 Dead in 5 Years, Forest and Energy Departments accuse each other', *The Odia Samaja,* Cuttack, 27 November 2012; S. Mishra and H. Mohanty, 'Wildlife, Power Authorities Lock Horns over Elephant Deaths', *Times of India*, Bhubaneswar, 9 October 2012; *The New Indian Express,* 'Forest, Energy Departments at Loggerheads', *Express News Service,* Bhubaneswar, 13 October 2012.

12. Letters no. 5129/1WL(H)/4/07 22 December 2012; no. 3647/4WL(G)4/2015, 7 May 2016 of the Office of the Principal CCF (Wildlife) & Chief Wildlife Warden, Bhubaneswar, Odisha.

13. E-mail with attachment titled 'Total Elephant Mortality in the country' received from Project Elephant Division, MoEF & CC, New Delhi, by the author at binoybehera@hotmail.com at 1730 hours and e-mail titled 'Death due to various reasons in the country' with attachments received at 1641 hours 4 March 2016.

14. Letter no NoJ-110 13/4112006-IA.II(I). Government of India Ministry of Environment & Forests, 27 June 2013, Available at www.moef.nic.in>assets>mef-letter-2013.

15. Available at http.hummingbirdmedia.org/wp-content/uploads/2016/05/African-elephant-geographical-range.jpg.

16. S. S. Bist, et al., 'The Domesticated Asian Elephants in India', available at http://www.fao.org/docrep/005/ad031e/ad031e0g.htm; as the 2016 census figures have not been finalised by Project Elephant yet, figures depicted are based on 2012 census.

17. *GKToday,* 'Elephant Census/Current Affairs Today', 17 August 2017, available at https//currentaffairs.gktoday.in; *Livemint,* 'India's Elephant Population Decreases by 10% to 27,312', available at www.livemint.com>Politics>Policy.

18. B. Scriber, '10 000 Elephants Killed by Poachers in Just Three Years, Landmark Analysis Finds', *National Geographic,* available at news.nationalgeographic.com/news/2014/08/140818-elephants-africa-poaching-cites-census/.

19. P. G. Allen, 'Great Elephant Census Final Results: The Final Report', available at www.greatelephantcensus.com.

20. J. Deb, 'Rhino-poaching Helps Fund Northeast India Rebel Group: Officials', 21 June 2016, available at www.benarnews.org/english/news/bengali/rhino-poacher-06212016164134.html.

21. S.G. Kashyap, 'Kaziranga Rhino Killed During Royal Couple Tour', *Indian Express,* Guwahati, 15 April 2016, available at http://indianexpress.com/article/india/india-news-india/hours-after-royal-couples-visit-rhino-killed-by-poachers-at-kaziranga-national-park/.

22. Viju B. and P. Oppilli/*TNN*. '100 Elephants Killed in 2 Years across South', *Times of India*, 13 August 2015.

23. Available at https://www.myind.net/behind-curtains-red-corridor; https://en.wikipedia.org/wiki/Insurgency_in_Northeast_India; http://www.thehindu.com/multimedia/archive/02945/Left_wing_extremis_2945018a.jpg.

24. *The New Indian Express*, 'In Move against Black Market, WWF Asks Asian Nations to Shut Tiger Farms', CAT-ATONIC, 29 July 2016.

25. *India Today*, 'Big Cat in Danger: International Poaching Mafia Gunning for India's Tigers', New Delhi, 7 August 2016, available at http://indiatoday.in/story/big-cat-in-danger-again-tiger-headcount-increases-so-does-poaching/1/734093.html; *Times of India*, '74 Tiger Deaths since January, available at http://timesofindia.indiatimes.com/home/environment/flora-fauna/74-tiger-deaths-since-January-drive-home-poaching-other-threats/articleshow/52965875.cms#; *Ommcom News*, 'India Loses 100th Tiger in 2016', *Indo–Asian News Service*, New Delhi, 13 October 2016, available at http://ommcomnews.com/public/india-loses-100th-tiger-in-2016.

26. V. R. Ramanath/*TNN*, 'Low Prey Base Big Worry for Simlipal', *Times of India*, Bhubaneswar, 18 September 2013.

27. *PTI*, 'Elephant Deaths Rampant in Odisha', *The Hindu*, Bhubaneswar, 31 October 2012; *The Samaja*, '49 Elephants Hunted with Electric Shocks—300 Dead in 5 Years, Forest and Energy Departments Accuse Each Other', Bhubaneswar, 27 November 2012; S. Mishra and H. Mohanty, 'Wildlife, Power Authorities Lock Horns over Elephant Deaths', *Times of India*, Bhubaneswar, 9 October 2012; *The New Indian Express*, 'Forest, Energy Depts. at Loggerheads', Bhubaneswar, 13 October 2012.

28. *The New Indian Express*, 'Too Many Die this Year and Blame Game begins yet Again', Bhubaneswar, 31 December 2012; D. Mohanty, 'Orissa Elephant Deaths: Panel Wants Speed Curbs for Trains', *The Indian Express*, Bhubaneswar, 6 May 2013; *The New Indian Express*, 'Jumbo Deaths: Rail Board Not to Restrict Train Speed', *Express News Service*, Bhubaneswar, 7 June 2013.

29. L. Neme, N. Kalron, and A. Crosta, 'Terrorism and the Ivory Trade', *The Los Angeles Times*, 14 October 2013; *National Geographic*, 'Al Shabab and the Human Toll of the Illegal Ivory Trade', 3 October 2013; Nil Kalron and Andrea Crosta, 'Africa's White Gold of Jihad: al Shabab and Conflict Ivory', available at https://elephantleague.org/project/africas-white-gold-of-jihad-al-shabaab-and-conflict-ivory/.

30. D. Vergano, 'Illegal Wildlife Trade Threatens International Security', *USA Today*, 24 June 2013, available at http://www.usatoday.com/story/news/world/2013/06/24/elephant-rhino-security.

Bibliography

Barbara, T. and A. Lister. 2000. *Elephants (Nature Fact Files)*. United Kingdom: Southwater.

Barua, M. 2010. 'Whose issue? Representations of human–elephant conflict in Indian and international media.' *Science Communication* 32 (1): 55–75.

Barua, M., J. Tamuly and R. A. Ahmed. 2010. 'Mutiny or clear sailing? Examining the role of the asian elephant as a flagship species.' *Human Dimensions of Wildlife* 15(2): 145–60.

Bhubaneswar, B. 2010. Chapter 4. (Source cannot be retrieved). In *Call of the Village, 115–16*. New Delhi: Anamika Publishers.

Bist, S. S. 2006. 'Elephant conservation in India—An overview.' *Gajah* 25: 27–35.

Chadwick, D. H. 1994. *The Fate of the Elephant*. San Francisco: Sierra Club Books.

Choudhury, A. et al. 2008. 'Elephas maximus, IUCN Red List of Threatened Species.' Version 2012.2, International Union for Conservation of Nature.

Choudhury, A and V. Menon. 2006. 'Conservation of the Asian Elephants in north east India.' *Gajah* 25: 47–60.

Choudhury, A. U. 1992. 'Trunk routes.' *WWF-Quarterly* 3(1): 14.

Chowdury, S. 2006. 'Conservation of Asian elephant in central India.' *Gajah* 25: 37–46.

Clutton-Brock, J. 1987. 'A natural history of domesticated mammals.' In *Natural History*, 208. London: British Museum.

Cosson M. J. 1997. *The Elephant's Ancestors*. Perfection Learning Corporation.

Cynthia, M. 1988. *Elephant Memories—Thirteen Years in the Life of an Elephant Family*. Chicago: University of Chicago Press.

Daniel, J. C., ed. 1980. 'The status of the Asian elephant in Indian sub-continent.' IUCN/SSC Report. Bombay: Bombay Natural History Society.

de Silva, S., A. D. G. Ranjeewa, and S. Kryazhimskiy. 2011. 'The dynamics of social networks among female Asian elephants.' *BMC Ecology* 11: 17. DOI:10.1186/1472-6785-11-17.

de Silva, S. and G. Wittemyer. 2012. 'A comparison of social organization in Asian Elephants and African Savannah Elephants.' *International Journal of Primatology* 33:1125-1141. DOI:10.1007/s10764-011-9564-1.

Desai A. 1997. *The Indian Elephant—Endangered in the Land of Lord Ganesha*. Vigyan Prasar.

Desai, A. A. 1995. 'Studies on population Ecology and Behaviour.' In *Ecology of the Asian Elephant, Final Report (1987–1992)*, 5–19. Bombay: Bombay Natural History Society.

D. K., Lahiri Choudhury. 1999. *The Great Indian Elephant Book*. New Delhi: Oxford University Press.

Elefantasia. 2008. 'Assist Us', 1 January.

Eltringham, S. K. and D. Ward. 1997. *The Illustrated Encyclopedia of Elephants—From Their Origins and Evolution to Their Ceremonial and Working Relationship with Man*. Salamander Books Limited.

Fernando, P et al. 2003. 'DNA analysis indicates that Asian Elephants are native to Borneo and are therefore a high priority for Conservation.' *PLOS Biology* 1-110-115.

Goswami, V. R., M. D. Madhusudan, and K. U. Karanth. 2007. 'Application of photographic capture-recapture modelling to estimate demographic parameters for male Asian elephants.' *Animal Conservation* 10: 391–399.

Gubbi, S. 2010. 'Are conservation funds degrading Wildlife habitats?' *Economic and Political Weekly* 45: 22–25.

Haynes, G. 1993. *Mammoths, Mastodons, and Elephants: Biology, Behaviour, and the Fossil Record*. Cambridge: University Press.

Heffner, R. and H. Heffner. 1980. 'Hearing in the elephant (Elephas maximus).' *Science* 208 (4443): 518–520. DOI:10.1126/science.7367876.

Hirschmann, K. 2005. *Elephants (Attack)*. Kid Haven Press.

Ian, R. 2001. *The Elephant Book*. Walker Books Limited.

Jerdon, T. C. 1874. *The Mammals of India*. London: John Weldon.

Karanth, K. U. and J. D. Nichols. 1998. 'Estimation of tiger densities in India using photographic captures and recaptures.' *Ecology* 79 (8): 2852–2862.

Kumar, A. and V. Menon. 2006. 'Ivory tower sustainability: An examination of the ivory trade.' In *Gaining Ground: 'In Pursuit of Ecological Sustainability'*, edited by D. M. Lavigne, 129–139. Canada: International Fund for Animal Welfare.

Lydekker, R. 1894. *The Royal Natural History, Volume 2*. London: Frederick Warne and Co.

Mariappa D. 1986. *Anatomy and Histology of the Indian Elephant*. Indira Publishing House.

Martin, E. and D. Stiles. 2003. *The Ivory Markets of East Asia*. Nairobi and London: Save the Elephants.

Menon, V. and M. Sakamoto. 1998[2002]. 'Analysis of the amended management system of domestic ivory trade in Japan,' unpublished report, Japan Wildlife Conservation Society, Tokyo, Japan.

Menon, V., R. Sukumar, and A. Kumar. 1997. *A God in Distress: Threats of Poaching and the Ivory Trade to the Asian Elephant in India*. Bangalore: Asian Elephant Research and Conservation Centre.

Menon, V. et al. 2005. 'Rights of passage: Elephant corridors of India.' *Wildlife Trust of India*, Conservation series No-3.

Menon, V. 2002. *Tusker: The Story of the Asian Elephant*. New Delhi: Penguin.

Miall, L. C. and F. Greenwood. 1878. *Anatomy of the Indian Elephant*. London: Macmillan.

Nīlakaṇṭha (of Rajamangalam). 1985[1931]. *The Elephant-Lore of the Hindus: The Elephant-Sport (Matanga-lila)*. New Delhi: Motilal Banarsidass.

Olivier, R. C. D. 1978. 'On the ecology of the Asian elephant.' Ph.D. thesis, University of Cambridge, UK.

Patterson, J. H. 1907. *The Man-eaters of Tsavo.* Macmillan and Co. Limited.

Payne, K. 1998. *Silent Thunder.* Simon & Schuster.

Pradhan, N. M. B. et al. 2008. 'Feeding ecology of two endangered sympatric megaherbivores—Asian elephant Elephas maximus and greater one-horned rhinoceros Rhinoceros unicornis in lowland Nepal.' *Wildlife Biology* 14: 147–154.

Price, S. 1997, 'Valuing elephants: The voice for conservation.' *Swara* 20(3): 29–30.

Raman, S. 2003. *The Living Elephants: Evolutionary Ecology, Behaviour and Conservation.* New York: Oxford University Press.

Rangarajan, M. 2001. 'The forest and the field in ancient India.' In *India's Wildlife History.* New Delhi: Permanent Black.

Rangarajan, M. et al. 2010. *Securing the Future for Elephants in India.* Report of the Elephant Task Force. New Delhi: Ministry of Environment and Forests.

Rodgers, W. A. and H. S. Panwar. 1985. 'Planning a wildlife protected area network in India Volume 1 and 2.' Dehradun: Wildlife Institute of India.

Roy, M., N. Baskaran, and R. Sukumar. 2009. 'The death of jumbos on railway tracks in northern West Bengal.' *Gajah* 31: 36–39.

Sakamoto, M. 1999. *Analysis of the Amended Management System of Domestic Ivory Trade in Japan.* Japan Wildlife Conservation Society, Tokyo, Japan.

Samansiri, A. P. and D. K. Weerakoon. 2007. 'Feeding behaviour of Asian Elephants in the north western region of Sri Lanka.' *Gajah* 27: 27–34.

Santiapillai, C. and Peter Jackson. 1990. *The Asian Elephant—An Action Plan for Its Conservation.* IUCN/SSC Asian Elephant Specialist Group.

Santiapillai, C. 1987. *Action Plan for Asian Elephant Conservation: A Country by Country Analysis—a Compilation.* World Wide Fund for Nature, Indonesia.

Sar, C. K. and D. K. Lahiri-Choudhury. 2002. 'A checklist of elephant movement paths/ corridors in Mahanadi catchment, Orissa.' *The Indian Forester* 128(2):235–242.

Saragusty, J. et al. 2009. 'Skewed birth sex ratio and premature mortality in Elephants.' *Animal Reproduction Science* 115(1–4): 247–254.

Shoshani, J. and J. F. Eisenberg. 1982. 'Elephas maximus.' *Mammalian Species* 182: 1–8. DOI:10.2307/3504045. JSTOR 3504045.

Shoshani, J. 2006. 'Taxonomy, classification, history, and evolution of elephants'. In *Taxonomy, Classification, and Evolution of Elephants,* edited by M. E. Fowler and S. K. Mikota, 3–14. Wiley-Blackwell.

Singh, A. K., R. R. Singh, and S. Chowdhury. 2002. 'Human-elephant conflicts in changed landscapes of south West Bengal, India.' *The Indian Forester* 128 (10): 1119–1132.

Singh, L. A. K. 1989. 'Elephant in Orissa distribution, status and management issues', paper presented at workshop on elephant issues, Dehradun.

Spinage, C. A. 1994. 'Elephants.' London: T&AD Poyser.

Stephen, A. 2004. *Elephas Maximus—A Portrait of the Indian Elephant.* United States: Houghton Mifflin Harcourt.

Stiles, D. 2009. *The Elephant and Ivory Trade in Thailand.* TRAFFIC Southeast Asia, Petaling Jaya, Selangor, Malaysia.

————. 1989. *The Asian Elephant—Ecology and Management.* Cambridge: Cambridge University Press.

Sukumar, R. and P. S. Easa. 2006. 'Elephant conservation in South India: Issues and recommendations.' *Gajah* 25: 47–60.

Sakamoto. 1999. Nishihara. 2003. 'What's wrong with selling southern African ivory to Japan?' *Wildlife Conservation Society Magazine.*

Swain, D. and S. K. Patnaik. 2002. 'Elephants of Orissa: Conservation issues and management options.' *The Indian Forester* 128(2): 145–154.

Talukdar, B. N. 2009. 'Elephants in Assam.' Wildlife Wing, Assam Forest Department.

Tiwari, S. K. et al. 2005. 'Elephant corridors of central India.' In *Rights of Passage: Elephant Corridors of India,* edited by V. Menon et al., 71–118. Wildlife Trust of India Conservation series No-3.

Vidya T. N. C. and R. Sukumar. 2005. 'Social organization of the Asian elephant (Elephas maximus) in southern India inferred from microsatellite DNA.' *Journal of Ethology* 23:205–210.

Vigne, L. and E. Martin. 2002. 'Myanmar's ivory trade threatens wild elephants.' *Gajah* 21: 85–86.

Index